EINSTEIN AND THE QUANTUM

EINSTEIN
AND THE QUANTUM

THE QUEST OF THE VALIANT SWABIAN

A. DOUGLAS STONE

PRINCETON UNIVERSITY PRESS

PRINCETON AND OXFORD

Copyright © 2013 by Princeton University Press

Published by Princeton University Press, 41 William Street, Princeton, New Jersey 08540

In the United Kingdom: Princeton University Press, 6 Oxford Street, Woodstock, Oxfordshire OX20 1TW

press.princeton.edu

Jacket photograph: Albert Einstein reading. Courtesy of The Hebrew University of Jerusalem / Corbis Historical Images. Personality rights of ALBERT EINSTEIN are used with permission of The Hebrew University of Jerusalem. Represented exclusively by GreenLight.

Library of Congress Cataloging-in-Publication Data

Stone, A. Douglas, 1954–

    Einstein and the quantum : the quest of the valiant Swabian / A. Douglas Stone.

       pages   cm

    Includes bibliographical references and index.

    ISBN 978-0-691- 13968-5 (hardback)

    1. Einstein, Albert, 1879–1955. 2. Physicists—Biography. 3. Quantum theory. 4. Science—History.

I. Title.

    QC16.E5S76   2013

    530.12—DC23   2013013162

British Library Cataloging-in-Publication Data is available

This book has been composed in Verdigris MVB

Printed on acid-free paper. ∞

Printed in the United States of America

10 9 8 7 6 5 4 3 2

This book is dedicated to my father, Alan, who has been my intellectual inspiration, and to my wife, Mary, who has been my emotional inspiration.

Science as something already in existence, already completed, is the most objective, most impersonal thing that we humans know. Science as something coming into being, as a goal, however, is just as subjectively, psychologically conditioned, as all other human endeavors.

—ALBERT EINSTEIN, 1932

CONTENTS

ACKNOWLEDGMENTS

This project began after I gave several public lectures at Aspen and at Yale on Einstein, Planck, and the beginning of quantum theory, when it became clear that most of this story was completely unknown both to the interested layman and to most working physicists. While several eminent historians of science, T. S. Kuhn, Martin Klein, Abraham Pais, and John Stachel for example, have written excellent but relatively technical works analyzing various facets of Einstein's work on quantum theory, no book for the general reader had attempted to synthesize all this into a complete picture. I have tried to fill that void with this book, while making it a fun read along the way. The book is based on the Collected Papers of Albert Einstein and the large body of outstanding historiography that has been produced on the history of quantum theory, blended with material from a number of biographies of Einstein, with a particular debt to the recent ones by Albrecht Folsing and Walter Isaacson. While I chose not to footnote quotations in the text, all their sources are identified in extensive notes at the back of the book.

I want to thank the late Martin Klein for his encouragement at the very early stages of this project, and Walter Isaacson for his generous advice and assistance, which was so important to a first-time author. I am very grateful to my editor, Ingrid Gnerlich, for her critical reading of the manuscript and useful guidance, and to Deborah Chasman, who made key suggestions for improving my initial draft. I also want to thank Samantha Hasey and Eric Henney at Princeton University Press, who helped with the final stages of preparation for publication. Barbara Wollf at the Albert Einstein Archive of the Hebrew University in Jerusalem was very generous with her advice and experience relating to the copyright permissions I was seeking, and Andy Shimp helped me navigate the library systems at Yale and retrieve difficult-to-find items. Both my father, Alan Stone, and my wife, Mary Schwab Stone, read the work with a keen eye and helped me immeasurably, not the least in keeping up my enthusiasm for the project. My son, Will Stone, found time between his journalistic pursuits to work as my editorial assistant in assembling the final version of the manuscript.

EINSTEIN AND THE QUANTUM

A HUNDRED TIMES MORE THAN RELATIVITY THEORY

"Let's see if Einstein can solve our problem." This was not an idea I had ever entertained, much less verbalized, during my previous twenty-six years doing research in quantum physics. Physicists don't read the works of the great masters of earlier generations. We learn physics from weighty textbooks in which the ideas are stated with cold-blooded logical inevitability, and the history that *is* mentioned is sanitized to eliminate the passions, egos, and human frailties of the great "natural philosophers." After all, since physical science (we believe) is a cumulative discipline, why shouldn't we downplay or even censor the missteps and misunderstandings of our predecessors? It is daunting enough to attempt to master and then extend the most complex concepts produced by the human mind, such as the bizarre description of the atomic world provided by quantum theory. Wouldn't telling the real human history of discovery just confuse people?

Thus, while I had studied history and philosophy of science avidly as an undergraduate, I had not read a single word written by Einstein during my actual career as a research physicist. I was of course aware that Einstein had contributed to the subject of quantum physics. Even freshman physics students learn that Einstein explained the photoelectric effect and said something fundamental about the quantized nature of light. And both atomic and solid-state physics (my specialty) have specific equations of quantum theory named for Einstein. So clearly the guy did *something* important in the subject. But the most familiar fact about Einstein and quantum mechanics is that *he just didn't like it.* He refused to use the theory in its final form. And troubled by the fundamental indeterminism of quantum mechanics, he famously dismissed its worldview with the phrase "God does not play dice."

Despite its esoteric-sounding name, quantum mechanics represents arguably the greatest achievement of human understanding of nature. By the end of the nineteenth century progress in physical science was stymied by the most basic problem: what are the fundamental constituents of matter, and how do they work? The existence of atoms was fairly well established, but they were clearly much too small to be observed in any direct manner. Hints were emerging from indirect probes that the microscopic world did not obey the settled laws of macroscopic Newtonian physics; but would scientists ever be able to understand and predict the properties of objects and forces so far from our everyday experience? For decades the answer was in doubt, until a theory emerged, a theory that has now withstood almost a century of tests and extensions. That theory has wrung human knowledge from the deep interior of the atomic nucleus and from the vacuum of intergalactic space. It is the theory that most physicists use every day in their work. This is the theory that Einstein rejected. Thus most physicists think of Einstein as playing a significant but still secondary role in this intellectual triumph.

I might have continued with this conventional view of Einstein and quantum physics for my entire career, if not for a coincidental intersection of my own research with that of the great man. I am interested in quantum systems, which if they were not microscopic but were scaled up in size to everyday proportions, would behave "chaotically." In physics this is a technical term; it means that very small differences in the initial state of a system lead to large differences in the final state, similar to the way a pencil, momentarily balanced on its point, will fall to the left or right when nudged by the smallest puff of air. I was searching (with one of my PhD students) for a good explanation of the difficulty that arises when mixing this sort of unstable situation with quantum theory. I recalled hearing that Einstein had written something related to this in 1917 and, almost as a lark, I suggested that we see if this work were relevant to our task.

Well the joke was on us. When we finally got our hands on the paper, we quickly realized that Einstein had put his finger on the essence of the problem and had delineated when it has a solution, *before* the invention of the modern quantum theory. Moreover, Einstein wrote with great lucidity about the subject, so that it seemed as if he were speaking directly to us, a century later. There was nothing dated or quaint about the analysis. For the first time in a long while, I found myself thinking, "Wow, this man really was a genius."

This experience piqued my interest in the actual history of Einstein and quantum theory, and as I delved into the subject I came to a stunning realization. It was Einstein who had introduced almost all the revolutionary ideas underlying quantum theory, and who saw first what these ideas meant. His ultimate rejection of quantum theory was akin to Dr. Frankenstein's shunning of the monster he had originally created for the betterment of mankind. Had Einstein not done so, in all likelihood *he* would be seen as the father of the modern theory.

This is not a view that one could extract from any of the popular biographies of Einstein, where the focus is always on his development of relativity theory. Nonetheless, I discovered to my surprise that, for much of Einstein's scientifically productive career, he was obsessed with solving the problems of quantum theory, not relativity theory. He commented to his friend the Nobel laureate Otto Stern, "I have thought a hundred times as much about the quantum problems as I have about Relativity Theory."

It is crucial to understand that while relativity theory is an important part of modern physics, for most of us quantum mechanics *is* the theory of everything. Quantum mechanics explains the periodic table of the elements, the nuclear reactions that power the sun, and the greenhouse effect that leads to global warming. The quantum theory of radiation and electrical conduction underlies all of modern information technology. Moreover, quantum mechanics has already subsumed part of relativity theory (the "special theory"). The goal of modern string theorists and their well-publicized "theory of everything" is to have quantum mechanics gobble up all of general relativity as well. Since quantum mechanics is the big kahuna, it behooves us to appreciate the role of Einstein in the "other" revolution of twentieth-century physics, the quantum one.

To understand Einstein's seminal role in this revolution, it is necessary to understand what had come before him. In this subject he had exactly one predecessor, the eminent German physicist Max Planck, of whom we will learn much below. Planck was the first major figure to recognize Einstein's seminal 1905 work on relativity theory, and he became Einstein's greatest champion in the world of science and one of his closest personal friends. But Einstein's work in the quantum theory—that was another matter. Sometimes it is easier to recognize the genius that doesn't paint in your own style. Planck had not worked on the problems that were solved by relativity theory, but he *had* worked on the quantum theory. In fact Planck, not Einstein, is universally regarded as its originator, based on his work on heat radiation in December 1900. Planck, a truly admirable man of science, indeed achieved something of incalculable significance as the new century began. But what it was, and what it meant, are not as clear as the textbooks maintain.

At the moment that Planck was making his historic advance the young Einstein, just graduated from the Zurich Polytechnic, was coming to a bitter realization: he was not wanted in the world of academic physics. Already engaged to his classmate Mileva Maric, as his travails, both practical and scientific, multiplied, he maintained a bold self-confidence. This was exemplified by a humorous nickname he chose for himself in his letters to Mileva: the "Valiant Swabian," after the swashbuckling crusader-knight invented by the Swabian romantic poet Ludwig Uhland. Einstein had just submitted his first research paper to the *Annalen der Physik*; it was on liquid interfaces and proposed a novel (but simplistic) picture of the forces between atoms. This would signify the beginning of his lifelong quest to understand the laws of physics on the atomic scale.

CHAPTER 1

"AN ACT OF DESPERATION"

On the evening of Friday, October 19, 1900, Max Planck, the world's leading expert on the science of heat, was experiencing a physicist's worst nightmare. Little more than a year earlier, he had staked his considerable reputation on a theory that purported to solve the outstanding problem of his field: the relationship between heat and light. Tonight at this meeting of the German Physical Society, the hall filled by the men who had been Planck's closest colleagues for over a decade, another scientist would announce publicly what Planck already knew—that the theory he had worked on for the past five years was almost certainly in error. This theory, which built on the work of his close friend Wilhelm Wien, was expressed in terms of a mathematical formula known as the Planck-Wien radiation law.

One of the scientists who had discovered the failure of the Planck-Wien law, Ferdinand Kurlbaum, was scheduled to speak first that night. A friend and close colleague of Planck's, Kurlbaum had no plan to attack Planck's theory on mathematical or logical grounds. Planck after all was the world's greatest expert on this topic and universally respected for his deep understanding of thermodynamics (the physics of heat flow and energy). Kurlbaum would simply present the hard data he and his collaborator, Henrich Rubens, had painstakingly collected to test the predictions of the Planck-Wien theory. The data would show (to quote Richard Feynman) that "Nature had a different way of doing things."

If Planck had been an experimenter himself, like Rubens and Kurlbaum, his reputation would have been less in jeopardy on that night. But Planck was a new breed of physicist, a *theoretical* physicist, with no laboratory or instruments. The theorist's job was (and is) simply to predict and understand physical systems, from stars and planets to atoms and molecules, using mathematical deductions from known and accepted physical laws. Very rarely (experience tells us about twice a century) theorists may also successfully propose some amendments to the laws of physics; but mostly they are master craftsmen, whose reputation depends on how well they use their intellectual tools. There had of course been great theory-building physicists before Planck: Isaac Newton, James Clerk Maxwell, and Ludwig Boltzmann, to name three of relevance to our story, but only at the end of the nineteenth century had the division of labor been formally recognized by academe, and the *theoretical* physicist, who divined nature by thought alone, became a recognized species. When Planck had taken up his post at the University of Berlin in 1889, it was the only chair of theoretical physics in Germany, and one of only a handful in the world.

Because a theorist has no measurements to report and no inventions to demonstrate, he is judged solely on whether his theoretical predictions describe important phenomena and are confirmed by experiment. An experimenter can go into the lab and make a great discovery without necessarily knowing what he is looking for, sometimes without even recognizing the discovery when it is first found. Many a Nobel Prize has been awarded for just such serendipity. In short, a good experimentalist can also be lucky. A good theorist, on the other hand, has to be right. Experimentalists are playing poker; theorists are playing chess. Chess games are not lost by "bad luck." The problem for Planck that night was that he had made a serious error in the contest with Nature, which was being exposed by Kurlbaum just now as Planck waited for his turn to speak. He needed to come up with an endgame that would preserve, at least temporarily, his reputation as a theorist.

So what was the problem on which the estimable Professor Planck had stumbled? It was the deceptively simple question of how much a heated object glows. The great Scottish physicist James Clerk Maxwell had demonstrated in 1865 that visible light and radiant heat are different expressions of the same physical phenomenon—the propagation of electrical and magnetic energy through empty space at the speed of light. The difference between visible radiation (i.e., "light") and thermal radiation is only their wavelength. For light, that length is about one-half of a millionth of a meter; for thermal radiation it is twenty times larger, or about ten millionths of a meter (which is still about eight times smaller than the width of a human hair). Such radiation arises when energy, originally stored in atoms (matter), is emitted; it can then be transmitted as an electromagnetic wave over large distances and be reabsorbed by matter.[1] In any enclosed space this happens over and over until the electromagnetic (EM) radiation and the matter share the energy in a balanced manner (they are "in equilibrium").

Thus matter is continually emitting and absorbing radiation—all objects are glowing, whether we can see their radiation or not. What determines *if* we can see it is the temperature of the object; at room temperature objects glow with primarily thermal (infrared) radiation, a wavelength that our eyes can't see (except with "night-vision" goggles). The red glow of heated metal appears when the metal becomes hot enough to emit just a little of its EM radiation as visible light. The surface of the sun, which is even hotter, emits most of its radiation at visible wavelengths.

The central problem of the physics of heat, the one that Max Planck had worked on for the past five years, was to understand and predict precisely, with a mathematical formula, the amount of electromagnetic energy coming out of an object of a given temperature at each wavelength. This formula is the law of thermal radiation; physicists had known such a formula should exist for over three decades, but finding the correct law and understanding it theoretically had frustrated the best minds of the era. Einstein himself commented somewhat later, "It would be edifying if the brain matter sacrificed by

---

[1] Electromagnetic radiation is not the only way heat is transmitted over distances; quite commonly a hot body (e.g., a heating coil in your stove) directly heats the air, which, as it moves around ("convects"), comes in contact with other bodies and heats them up.

theoretical physicists on the altar of this universal [law] could be put on the scales; and there is no end in sight to this cruel sacrifice!" In 1899, roughly a year and a half earlier, Planck thought he had found the answer, and had proudly announced his conclusions to the very same audience he was scheduled to address this evening. At that earlier meeting he had derived mathematically the equation that generated a universal curve, or graph, with temperature on the horizontal axis and EM energy on the vertical.

The current speaker, Kurlbaum, was presenting his and Rubens's measurements of just this curve, as Planck waited in the audience to respond. The data made a neat straight line, showing that the infrared energy radiated by an object increased proportional to the increasing temperature. On the same graph the prediction of the Planck-Wien law was plotted, giving a rainbow-shaped curve with not even a passing resemblance to the actual measured data points.

Planck had known that this moment was coming. Rubens was a personal friend, and he and his wife had visited Planck twelve days earlier for Sunday lunch. As physicists are wont, Rubens began talking shop and informed Planck that the law of thermal radiation that Planck had defended ardently for the past two years was badly out of agreement with their new data, which instead showed an intriguing linear variation with temperature. It was on this rather dramatic failure of his theory that Herr Planck would soon be asked to "comment." Thus the impending discussion showed every sign of being exceedingly awkward.

Planck was no longer a young man, although he was famously vigorous, and would climb mountains well into his seventies. At forty-two his hair was receding above his piercing eyes, and it sometimes pushed straight upward in an unruly shock. He had the bushy handlebar mustache sported by many of his Prussian colleagues and was dressed neatly in the academic style: white shirt with high collar, black bow tie and jacket, and pince-nez glasses. As a young man he had gone into science for the most idealistic of reasons: "my decision to devote myself to science was a direct result of the discovery that . . . pure reasoning can enable man to gain an insight into the mechanism

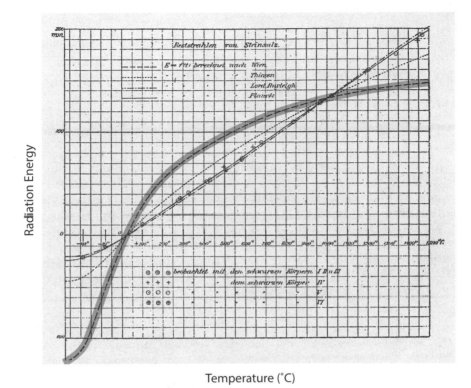

FIGURE 1.1. Original data showing measurements of blackbody radiation energy (vertical axis) as the temperature (horizontal axis) is varied while the frequency is fixed compared to different theories for the Radiation Law. The data points are represented by various types of symbols, with the different types of dashed lines represent different theories. The curve with the larger dashes outlined in gray represents Wien's Law which disagrees strongly with the data. The small dashes represent an empirical formula of no historical importance. The dash-dotted line is the Raleigh-Jeans Law which works rather well for the long wavelength (low frequency) radiation measured in this experiment (but which fails in other experiments). The solid line is Planck's Law, which fits the best and also works at higher frequencies. The graph is from 1901, shortly after Planck proposed his law; in October of 1900 he still believed that the Wien Law was correct. More details on the Radiation Laws are given in appendix 2.

of [natural laws]. . . . In this connection it is of paramount importance that the outside world is something independent from man, something absolute, and the quest for these laws . . . appeared to me as the most sublime scientific pursuit in life." Early in his academic career he had been attracted to the science of heat, thermodynamics, since it is based on two absolute laws. The First Law states that heat is a form of energy, and the Second Law governs the flow of heat and

FIGURE 1.2. Max Planck in 1906, six years after he initiated the quantum revolution. Courtsey Archiv der Max-Planck-Gesellschaft, Berlin-Dahlem.

the possibility of converting heat energy to do useful work, as in a steam engine. The Second Law employs the mysterious concept of entropy (roughly speaking, the amount of disorder in a physical system), and Planck had based his career on the interpretation and applications of this profound notion. That was why he was now in trouble.

Planck had not presented the Planck-Wien law of thermal radiation as a conjecture, based on provisional assumptions that he might revise. Quite the contrary. Little more than a year before, standing in front of the very same group of physicists, he had "proved" to them that this law followed from no other assumption than the Second Law of thermodynamics. With crushing certainty he had stated, "the limits of validity of this law coincide with those of the Second Law of Thermodynamics." This was the heavy artillery; the Planck-Wien law was supposed to be as solid as the Second Law itself! Einstein, also an admirer of thermodynamics, has said it is "the only physical theory of universal content which, within the framework of the applicability of its basic concepts, I am convinced will never be overthrown" (and, so far, he has been right). So if Planck, the world's expert, said that he had derived the law of thermal radiation directly from the Second Law, the case should have been closed. Unfortunately for Planck, the data disagreed.

Thus, when Planck stepped to the podium that night, his aim was not scientific revolution but damage control. Nonetheless, he was a truth seeker; he was not willing to run away from unpleasant facts. Later he scorned the English theorist James Jeans for just such behavior: "He is the model of the theorist as he should *not* be. . . , [because he believes] so much the worse for the facts if they don't fit." Planck stood up and faced the music: "The interesting results of long wavelength spectral energy measurements . . . confirm the statement . . . that Wien's energy distribution law is not generally valid. . . . Since I myself *even in this Society* have expressed the opinion that Wien's law must be necessarily true, I may perhaps be permitted to explain briefly the relationship between the . . . theory developed by me and the experimental data."

The "relationship" between them of course is that the Planck-Wien theory is wrong; Planck could not quite bring himself to say *that* in his remarks. But he did identify a weak point in his earlier arguments and admitted that the Second Law of thermodynamics does not have enough power, on its own, to answer the question. There had to be some further new principle involved. Having lost his guideposts for the journey, but being under such intense pressure to come up with an answer, Planck did something highly uncharacteristic. Planck was not a man to leap impulsively into the unknown; by his own description he was "by nature . . . peacefully inclined, and reject[s] all doubtful adventures." Nonetheless, on that October night he had decided to wing it. What followed was the most fateful improvisation in the history of science.

Planck had been fortunate that his friend Rubens had given him warning of the failure of his theory. Moreover Rubens's data provided a huge clue to what was wrong. Earlier experiments had shown that the Plank-Wien law worked very well for *visible* EM radiation emitted by very hot bodies, that is, for the shorter wavelengths. The new experiments of Rubens and Kurlbaum showed not only that the law failed for the longer, infrared wavelengths emitted by less hot objects, but also showed exactly *how* it failed. That nice, straight line in the data told Planck that at long wavelengths, contrary to the prediction of the Planck-Wien law, the radiation energy must be proportional to temperature. To Planck the challenge was similar to filling in a line in a crossword puzzle for which the end of the word was known, and now someone had filled in the first letter for him, telling him his original guess was wrong. With a little inspired mathematical insight on the very Sunday night, twelve days earlier, that Rubens had warned him of the problem, Planck had *guessed* the correct mathematical formula for the law of thermal radiation. Now, at the meeting, he unveiled his new formula, soon to become immortalized as the Planck radiation law.[2] Moreover he took the liberty of sketching how his new law compared to the Rubens-Kurlbaum data; it produced a line perfectly matching the data points. He concluded, "I should therefore be permitted to draw your attention to this new formula, which I consider to be the simplest possible, apart from Wien's expression."

With this great leap of intuition Planck had achieved a draw, but not a victory. Theorists are not supposed to just *guess* the correct formulas to describe data; they are supposed to *derive* these formulas from the fundamental laws of physics, which at the time were Newton's laws of mechanics and of gravity, Maxwell's electromagnetic theory, and the laws of thermodynamics. For Planck's new law to be anything more than a "curiosity" (as he himself put it), he would have to connect it to the more general laws of physics. As Planck himself said, "even if the absolutely precise validity of the radiation formula is taken for granted, so long as it had merely the standing of a law disclosed by a lucky intuition, it could not be expected to possess more than a formal significance. For this reason, on the very day when I formulated this law, I began to devote myself to the task of investing it with true physical meaning."

The details of Planck's new reasoning will be fully explained later in our story. For the moment it suffices to say that after "some weeks of the most strenuous work" of his life, "some light came into the darkness," and Planck again went before the German Physical Society to justify his radiation law. In the course of that presentation, on December 14, 1900, he uttered two sentences of incalculable significance for humankind:

We consider, however—this is the most essential point of the whole calculation—[the energy] E to be composed of a very definite number

---

[2] Moreover, the old terminology referring to the incorrect law, the "Planck-Wien law," was quickly adjusted to simply the "Wien law," erasing from the physics canon the evidence of Planck's original error.

of equal parts and use thereto the constant of nature $h = 6.55 \times 10^{-27}$ erg-sec. This constant, multiplied by the frequency v . . . gives us the energy element, ε.

Planck showed that, from this assumption and the then-controversial statistical theory of atoms, his new law of thermal radiation followed. But with this cryptic phrase natural science had crossed a philosophical Rubicon: ultimately the exquisitely sharp Newtonian photograph of the natural world would fall out of focus, becoming blurred and uncertain. The even flow of natural processes would give way to an atomic world of sudden jumps and collapses. Light itself would become grainy, belying its wave properties, so brilliantly wrung from the nineteenth-century triumphs of Maxwell and others. And all who look to science to elucidate the universe would have to get used to a worldview that sanctioned "spooky action at a distance," the modern quantum view of reality.

Planck's insight was beyond brilliant; it was an act of genius. The new law he introduced will bear his name and will be used by scientists as long as there is technologically advanced human civilization. In fact, as far as we know, it may be in use right now in nonhuman civilizations.[3] The theory that arose from this insight, the quantum theory, is unquestionably the most important theoretical advance in physical science since Newton.

So by December 1900 Planck had changed everything in physics and chemistry. The only problem was he didn't realize it. Planck was still recovering from his near-death experience as a reputable theorist. He later described his arrival at the quantum hypothesis as "an act of desperation." Now he breathed a deep sigh of relief and put the "energy element" out of his mind: "I considered the [quantum hypothesis] a purely formal assumption, and I did not give it much thought except for this: that I had obtained a positive result under any circumstances and at whatever cost." And the entire physics community went along

[3] Surprisingly, the staid Planck had some things to say about extraterrestrials, as we will see in chapter 14.

with this denial, like a family with an unspoken agreement to never again discuss a traumatic event.

Although he didn't realize it, Planck had removed a foundation stone from the edifice of classical physics; it would take another twenty-five years for the entire structure to collapse. However, the immediate reaction was . . . nothing. For the next five years neither Planck nor any of the great physicists of the era took up the meaning and extension of Planck's ideas. Not the revered Hendrik Lorentz in Holland, nor the profound but impenetrable Ludwig Boltzmann in Vienna, nor any of Planck's close colleagues picked up the challenge. That was left to a twenty-five-year-old patent examiner and maverick theorist living in Bern, Switzerland. Like Planck, thermodynamics and statistical mechanics had been his first love as a physicist. In addition he had been fascinated from an early age by Maxwell's equations and EM radiation. Unlike Planck however, *he* had been rejected by academe and had no reputation to lose. He was on the verge of taking the leap that Planck and the other great physicists of the time had not even considered. He was about to give Planck's radiation law the most radical interpretation possible: that it implied the discontinuity of motion on the atomic scale. He would begin this uprising with a paper that he himself termed "revolutionary." His name was Albert Einstein.

CHAPTER 2

THE IMPUDENT SWABIAN

We do not know the exact moment when Heinrich Weber began to despise Albert Einstein. It definitely was not at first sight. Professor Weber was the head of the Physics Department of the Federal Institute of Technology, an up-and-coming engineering school in Zurich, Switzerland, now known worldwide as ETH Zurich. In 1895, when Weber and Einstein first met, the "Poly" (as it was called by the locals) had the immense advantage for the young Einstein that it did not require a high school diploma for admission. This was particularly pertinent for Einstein because he had rather recently and without the consent of his parents "excused himself" from the final two years of his well-regarded German high school (the Luitpold Gymnasium in Munich) on the basis of a nebulous medical condition, "neurasthenic exhaustion." In fact he had hated the school, and once his parents had left Munich for Italy for financial reasons he saw no reason to stick it out. In late December of 1894 the fifteen-year-old Einstein showed up on their doorstep in Milan and "assured them most resolutely" that he would self-school himself in order to qualify for admission to the Zurich Poly by the next fall.

Indeed Einstein was already an accomplished autodidact, having taught himself differential and integral calculus well ahead of the school curriculum; to qualify for the Poly he had taken the precaution of obtaining a letter of advanced mathematical achievement from his teacher in Munich. Armed with this certificate, he presented himself to Albin Herzog, principal of the Zurich Poly, in October of 1895, as a "prodigy" who should be allowed to take the entrance exams a full year and a half before he would attain the required minimum age. It was at this time that he encountered Professor Weber, a reserved and dignified scientist, who, while not a physicist of historic stature, was a respected experimental researcher in thermodynamics.

On the entrance exams Einstein confirmed the judgment of his mathematics teacher, performing brilliantly on the math and physics portions of the test. However, he was neither fond of nor talented in subjects requiring a great deal of memorization, so he failed the general sections of the exam, which covered subjects such as literature, French, and politics. He thus failed to gain admission to the Poly. Yet his strong showing in math and physics so impressed Weber that he invited Einstein, against regulations, to attend his own lectures for second-year physics students. But there was still the minor matter of qualifying for admission, which could not be met by auditing lectures in the subjects where Einstein already excelled. So, at the suggestion of Herzog, Einstein enrolled in the cantonal high school in nearby Aarau for an additional year of formal schooling. He thrived there, finishing first in the final exams and gaining automatic admission to the Poly in October of 1896.

It was then that his intense relationship with Weber began. Weber was the primary physics instructor, and Einstein took fifteen courses from him, ten in the classroom and five in the lab. He did well in all of them. His very first physics course was with Weber, who immediately impressed him. "Weber lectured on heat . . . with great mastery. I am looking forward from one of his lectures to the next," Einstein wrote in 1898 to his fellow student and future wife, Mileva Maric. Einstein had been fascinated with physics since he was a young boy, beginning with his experience of "wonder" at a compass that he received at age five, which revealed to him the existence of unseen forces. Initially at the Poly his youthful love of physics was nurtured, and he responded with a strong academic performance: at the end of his first two years he passed the intermediate diploma exam with an overall grade of 5.7 out of 6, placing him first in his class.

But a problem was emerging. Einstein was aware of the great advances in physical theory that had taken place in the previous twenty years, and specifically of the world-changing electromagnetic theory of Maxwell and the bold statistical theory of gases due to Boltzmann. In vain he awaited the appearance of these exciting ideas in his classroom. Weber was a conservative scientist and had no intention of teaching these recent and highly mathematical developments in his lectures. A fellow student remarked: "[Weber's] lectures were outstanding but . . . [modern developments] were simply ignored. At the conclusion of one's studies one was acquainted with the history of physics, but not with its present or future." In particular, "Einstein's hopes of learning something decisive about Maxwell's electromagnetic theory were not realized." When, in his very last semester, Einstein heard a lecture from the mathematician Hermann Minkowski on modern formulations of Newton's laws he remarked to a classmate, "this is the first lecture on mathematical [theoretical] physics we have heard at the Poly." This refusal to admit the existence of newer ideas in physics apparently revived in Einstein a long-standing characteristic of his personality, his disrespect for authority in the classroom.

In later life Einstein spoke many times about how regimented his early education had been, and how he had particularly disliked the German system, as exemplified by the Luitpold Gymnasium, which he had fled in 1894. But contemporaries of his, even those with the same Jewish background, did not recall this school as being oppressive. In fact there seems to have been something in Einstein's own manner that contributed to the conflicts he recalled; he had a knack for driving his teachers to distraction. At the gymnasium his frustrated Latin teacher, Joseph Degenhart, offered one of history's great erroneous predictions: that "[Einstein] would never get anywhere in life." When Einstein maintained that he had committed no offense to elicit such an opinion, Degenhart replied, "your mere presence here undermines the class's respect for me." After leaving the gymnasium and ending up at Aarau, despite his much greater affinity for this school, his tendency to be less than respectful to his teachers did not change. While on a field trip he was questioned by his geology teacher, Mühlberg, as to whether the strata they were observing ran upward or down; he replied, "It is pretty much the same to me which way they run, Professor." An Aarau classmate recalled the impression made by the young

Einstein. "A cold wind of skepticism was blowing [that suited] the impudent Swabian. . . . Sure of himself, his grey felt hat pushed back from his thick black hair, he strode energetically up and down in the rapid, I might almost say crazy tempo of a restless spirit which carries a whole world in itself. Nothing escaped the sharp gaze of his large brown eyes. Whoever approached him came immediately under the spell of his superior personality. A sarcastic curl of his lip did not encourage Philistines to fraternize with him. . . . his witty mockery pitilessly lashed any conceit or pose."

At Zurich, by his third year of study, the impudent Swabian had reemerged, much to the dissatisfaction of Professor Weber. Einstein began skipping classes and studying the modern works independently, and with "holy zeal," in his room or in cafés with friends. Weber noticed that, unlike the other students, Einstein always addressed him as "Herr Weber" and not as "Herr Professor," a brash gesture in the hierarchical Germanic universities of the time. While Einstein continued to do well in Weber's classes, his arrogance surfaced most clearly in the experimental laboratory class of Weber's colleague Jean Pernet. At the beginning of each laboratory class the students would be given a sheet of instructions, or "chit," on how to perform the required work. Einstein would ostentatiously fling the sheet into the wastebasket and proceed by his own methods. This infuriated Pernet, and not necessarily without good reason, as Einstein's explorations eventually caused an explosion in the lab resulting in an injury to his right hand requiring stitches, which appears to have upset him only because it prevented him from playing the violin for several weeks. Pernet seems to have sincerely misjudged Einstein's ability, telling him "there is no lack of eagerness or goodwill in your work, but a lack of capability." When Einstein protested that he felt he did have a talent for physics, Pernet answered curtly, "I only wanted to warn you in your own interests." Eventually Pernet did more than warn Einstein; he flunked him in the course and had him placed on academic probation.

Professor Weber of course was aware of all this and must also have sensed Einstein's loss of respect for him because of his backward-looking pedagogy. By Einstein's third year Weber had developed a particular dislike for the young man, and this boiled over when he confronted Einstein with the following critique: "You're a very clever boy, Einstein, an extremely clever boy, but you have one great fault: you'll never let yourself be told anything." Of course Weber was right on target, except for his choice of the word "fault."

By the end of his final year, in the spring of 1900, it was clear that Einstein had made an enemy of his former mentor. Previous to his final exams Einstein had to produce for Weber a diploma thesis. Weber had rejected an earlier proposal by Einstein for a study

FIGURE 2.1. Albert Einstein, "the impudent Swabian," age 17, at his graduation from the cantonal school in Aarau, Switzerland. ETH-Bibliothek Zurich, Image Archive.

of the electron theory of electrical current and heat flow, and Einstein had resigned himself to discussing a more mundane topic in heat conduction, in which he had little interest. Three days before his final exams, when Einstein must have been quite pressured to prepare for them, Weber made him recopy the entire diploma thesis because he had not submitted it on the "regulation paper." To top it off, Weber then gave the thesis the poor grade of 4.5 out of 6.

Einstein's falling out with Weber might not have had such major consequences for his immediate future if not for another imprudent decision he made. While the physics faculty of the Poly did not contain scholars of historic distinction, the mathematics faculty did: specifically the number and function theorist Adolf Hurwitz and the geometric number theorist Minkowski, whose lecture on mathematical physics Einstein had admired. Minkowski was later to reformulate Einstein's special theory of relativity as a description of four-dimensional space-time, becoming famous beyond science for his dramatic statement that "henceforth space by itself and time by itself are doomed to fade into mere shadows and only a kind of union of the two will preserve an independent existence." Einstein, with his firm commitment to theoretical work and his obvious talent for mathematics, could have impressed these important men with his promise, had he deigned to attend their advanced classes. But he decided that he knew all the math he needed for the physics he intended to do and focused on his independent study of contemporary physics. That the strongest student in his year (at the halfway point of the program) decided to skip the most challenging math courses did not go unnoticed. Years later Minkowski told the physicist (and future close friend of Einstein) Max Born that Einstein's success had come as a "tremendous surprise, for in his student days Einstein had been a lazy dog. He never bothered about mathematics at all." Thus by his graduation in June of 1900 Einstein had alienated all the professors whom he might reasonably have asked for the thing he really needed, a position as an academic assistant. This was the standard means for a young graduate to prepare for a career in research and teaching, and since it was very poorly paid, it was rather routine to obtain, particularly for a student of promise such as Einstein.

However it didn't work out that way for young Albert. First, cramming in all the material he had missed using the notes of his friend and classmate Marcel Grossmann did not compensate for the many lectures he had skipped. He produced a mediocre performance on the final exams, and, handicapped by the poor diploma grade he had already received from Weber, he posted the lowest combined grade point average of the four physics students who passed. (The fifth student taking the exams was Einstein's future wife and current amour, Mileva Maric; she did even worse and was not granted the degree.) The other three passing students (Grossman, Kollros, and Ehrat) were immediately granted posts as assistants to members of the math faculty. Professor Weber, who needed *two* assistants for the physics program, as his final insult, engaged two engineering graduates in preference to Einstein.

CHAPTER 3

THE GYPSY LIFE

"This Einstein will one day be a very great man." Although Einstein had finished college having made a poor impression on his professors, the opposite was true of his peers. Einstein already was a man of great charisma and charm, to go along with his penetrating intellect and deep understanding of science. The prediction of greatness came from Einstein's classmate Marcel Grossmann, only a few days after their first meeting in 1898. Grossmann, according to Einstein, "was a model student, close to his teachers"; he would graduate at the top of their class and within a few years become a mathematics professor himself at the Poly. He and Einstein formed a genuine friendship: "Once every week," Einstein recalled, "I would solemnly go with him to the Café Metropol on the Limmat Embankment and talk to him not only about our studies, but also about anything that might interest young people whose eyes were open." Besides providing Einstein with the invaluable class notes, Grossmann frequently had him as a visitor to his home. Grossmann came from an old and well-connected family near Zurich and would eventually save Einstein from his "vagabond existence" after graduation by arranging for him (through his father's connections) a job at the patent office in Bern.

But it was not only the important Grossmann family whom Einstein charmed; his sympathies were completely independent of class or status. The woman who ran his boardinghouse, Stephanie Markwalder, forgave him his repeated losing of the house key, as "his impulsive and upright nature . . . was so irresistible." To the delight of Frau Markwalder and her daughter Suzanne, he filled their house with joyful and passionate music-making, as well as vigorous discussions, puns, and witticisms. Suzanne, who often accompanied Albert's violin playing on the piano, found his joie de vivre infectious, as for example when she encountered Albert and Mileva descending the Uetliberg mountain near Zurich and he called out to her, "you must go all the way to the top. It's marvelous up there . . . it's covered with tiny feathers (hoar frost)."

His generosity extended in all directions. Once, when he showed up late to his regular meeting with Grossmann and comrades in the Café Metropol, he explained that the laundry woman at the boardinghouse had told him his violin playing made her work more agreeable, so he had stayed on to please her. On another occasion he came to the rescue of a biology student struggling to please the demanding professor in her physics lab course. After witnessing the student being dressed down by the professor, he offered to take her lab notebook home with him and returned it to her with results that found commendation from the difficult man. However, when she told him he was lucky to work for Weber instead, he couldn't agree: "What the one teaches is not right, but what the other teaches is not right either!" Another physics classmate, Jakob Ehrat, often sat with Einstein in class, and they became close. Ehrat was an anxious type, not nearly as facile with the material as Einstein; Einstein would help him maintain his composure before the final exams (in which he ended up doing better than Einstein). Einstein often visited him and his mother at their home in Zurich, and on one memorable visit the young Einstein showed up wearing the runner from his chest of drawers as an improvised scarf. Ehrat was amused by his eccentricity but did not fail to see the true character of the man. "I never saw a trace of pettiness, the slightest weakening in his courage for truth and in his refusal to compromise. His almost prophetic gift for justice, his inner strength and spontaneous feeling for beauty impressed me so much that I often dreamed of him long after life had separated us."

Not surprisingly Einstein's personal magnetism, artistic temperament, and striking appearance engendered romantic feelings as well as admiration in the opposite sex. A male student whom he taught after graduation wrote down the following somewhat analytical description, which nonetheless makes the point: "Einstein is 5 ft 9, broad shoulders and a slight stoop. Unusually broad short skull. Complexion a matte light brown. A garish moustache sprouts above his large sensual mouth. Nose rather aquiline, and soft deep brown eyes. The voice is compelling, vibrant like the tone of a cello." A friend of his second wife described his looks more succinctly: "He had the kind of male beauty that, especially at the beginning of the century, caused such havoc." Before coming to Zurich Einstein had already captured the heart of Marie Winteler, the sweet and naive daughter of his

FIGURE 3.1. Einstein, aged 19, as a student at the Poly, roughly at the time when he was becoming involved with his first wife, Mileva Maric. ETH-Bibliothek Zurich, Image Archive.

host family in Aarau. He broke off this dalliance soon after arriving at the Poly. It was not long before he had a new object of his interest, his fellow physics student and classmate Mileva Maric.

Maric was three years older than Einstein and was, not surprisingly, the only woman in the physics track at the Poly. It was only through the great efforts of her father, Milos Maric, a Serbian peasant who had married into the middle class, that she had managed to obtain a math and physics education, from which women were normally excluded. Having excelled at the Zagreb Classical Gymnasium, she had enrolled in Zurich, with every intention of pursuing a career in science. As the class was so small, she met Einstein immediately in the fall of 1896, and there seems to have been some instant chemistry between them, as there is evidence that her decision to leave Zurich for Heidelberg a year later was based on a need to get some distance from him and her emerging feelings for him. However she was far from the meek, simple sweetheart that Einstein had encountered in Marie Winteler; here is how she answered his first letter to Heidelberg:

> It's now been quite a while since I received your letter, and I would have replied immediately . . . but you said I should write to you someday when I happened to be bored. And I am very obedient and I waited and waited for boredom to set in; but so far my waiting has been in vain and I really don't know how to manage this; I could wait from here to eternity, but then you would be right to take me for a barbarian, and, again, if I write my conscience is not clear.

These are the words of a spirited, independent woman who could almost match Einstein's sarcastic wit and also his love of science. Physically she was not considered a great catch; a friend described her thus: "very smart and serious, small, delicate, brunette, ugly." She walked with a limp from a congenital hip defect and had endured tuberculosis growing up; she suffered from bouts of depression, something to which Einstein seemed immune even during these trying early years ("I am a cheerful fellow . . . and have no talent whatsoever for melancholic moods," he wrote in 1901, in the midst of his struggles with family and joblessness). All this meant nothing to Einstein at the time: he had found a soul mate, an outsider and rebel like him, who would join him on his journey of intellectual discovery. Maric returned to the Poly in February of 1898, and their relationship blossomed; a growing intimacy intertwined with a shared sense of discovery as Einstein began to confide in her his ideas that implied that important physical theories were wrong and had to be changed.

As early as March of 1899 Einstein wrote to Maric and told of showing her photograph to his mother and of the tense interchange that ensued (his mother's deep antipathy to Maric would plague the couple for many years). However, he ends the letter with an intriguing hint: "My broodings about radiation are starting to get on a somewhat firmer ground & I am curious whether something will come out of them." By the summer of that year he had begun addressing her with the affectionate nickname "Dolly" (Doxerl). In one letter, after explaining to her, complete with equations, why he believes the present electrodynamic theory cannot be correct, he ends with: "If only you would be again with me a little! We both understand each other's black souls so well & also drinking coffee & eating sausages, etc." (Among Einstein's inventions at that time was the suggestive form of etcetera.) The summer of 1900 brought twin disappointments: Einstein's lack of success in landing an assistant's position, and Maric's failure to pass the final exams and receive her degree. If anything this seemed to bring them closer together, and in July Einstein, during a visit home, announced his intention to marry Mileva, eliciting a "scene" of anger and disapproval from his mother.

Einstein was unmoved by this display, but as often happened, he found solace in reading physics papers, in this case the works of Gustav Kirchoff, the first physicist to understand the importance of the thermal radiation law. Both of Einstein's parents continued the barrage of disapproval of Maric over the summer, but Einstein held firm: "Mama and Papa are very phlegmatic people and have less stubbornness in their whole body than I have in my little finger." By the fall of 1900 it was clear that for the immediate future it would be Albert and Mileva against the world, both the world of bourgeois society and the world of the physics establishment. In October Einstein wrote his fiancée from Milan: "You too don't like the philistine life any longer, do you! He who tasted freedom cannot stand the chains any longer. How lucky I am to have found in you a creature who is my equal, who is as strong and independent as I am myself!"

It was not immediately obvious to Einstein how serious his plight was as he cheerily expected something to turn up any day. He even had the temerity to turn down an insurance job that his friend Ehrat had lined up for him, saying "One must shun such stultifying affairs." Initially he clung to the naive hope that he could land a job with the mathematician Hurwitz, whose courses he had avoided rather brazenly. When this didn't work out, he wrote to a friend: "Neither of us have gotten a job and we support ourselves by private lessons, when we can pick up some, which is still very questionable. Is this not a journeyman's or even a gypsy's life? But I believe we will be happy in it nonetheless."

CHAPTER 4

TWO PILLARS OF WISDOM

The man loved mysterious Nature as a lover loves his distant beloved. In his day there did not exist the dull specialization that stares with self-conceit through horn-rimmed glasses and destroys poetry.

—ALBERT EINSTEIN, ON MICHAEL FARADAY

"About Max Planck's studies on radiation, misgivings of a fundamental nature have arisen in my mind, so that I am reading his article with mixed feelings." So Einstein wrote to his Dolly from Milan in April of 1901, a scant four months after Planck's "act of desperation" in Berlin had saved his own reputation but failed to alert the physics community to the storm ahead. In the same letter Einstein ruefully admits, "soon I will have honored all physicists from the North Sea to the southern tip of Italy with my [job inquiry]." Emboldened by his first published article, which had appeared in the prestigious journal *Annalen Der Physik*, Einstein had sent a slew of postcards requesting an assistant's position to the well-known physicists and chemists of Europe. None of these missives bore fruit, and as far as we know few of them were even graced with a reply. Although Einstein was convinced that Weber was behind the rejections, Einstein's indifferent final academic record and his failure to receive the pro forma job offer from the Poly would likely have been enough.

Despite these disappointments he was scraping together a living through part-time jobs and private lessons and forging ahead with his independent thinking about the current state of theoretical physics. For much of this time he would be separated from his fiancée, but writing to her frequently. In his very next letter to Maric he continues discussing Planck: "Maybe his newest theory is more general. I intend to have a go at it." A little later in the letter he comments, "I have also somewhat changed my idea about the nature of the latent heat in solids, because my views on the nature of radiation have again sunk back into the sea of haziness. Perhaps the future will bring something more sensible." His last words were prescient; his views on radiation would emerge from haziness to enlarge the Planck radiation theory in a revolutionary manner, while the latent (or specific) heat of solids, a seemingly mundane topic, would provide Planck's theory with the radical physical interpretation that it currently lacked. But before this could occur, Einstein needed to plumb deeply into thermodynamics, Planck's specialty, and the newer atomistic discipline of statistical mechanics, which attempted to explain and extend the laws of thermodynamics. His main scientific motivation at the time was not to unravel the puzzles of relative motion. Einstein's famous insight, that resolving these puzzles would require a major reshaping of our conceptions of time and space, would not occur to him for four more years. Rather, his primary scientific focus from his student days was "to find facts which would attest to the existence of atoms of a definite size." Proving the existence of atoms and understanding the physical laws governing their behavior was the original quest of the Valiant Swabian.

The atomic world was the frontier of physics at the beginning of the twentieth century. The disciplines of what is now called classical physics had all developed without a need to delve too deeply into the question of the nature of the microscopic constituents of matter. That situation had now changed. If physics was going to progress, it would be essential to understand the fundamental origin of electromagnetic phenomena, of heat flow, of the properties of solids (e.g., electrical conductivity, thermal conduction and insulation, transparency, hardness), and the physical laws leading to chemical reactions. The answers to these questions would only be found by understanding the makeup of the atom and the physical interactions between atoms and molecules.

Modern physics had begun with the work of Sir Isaac Newton in the second half of the seventeenth century. He introduced a new paradigm for the motion of objects (masses) in space: first by the bold assertion that the natural state of motion of a solid body is to move at a constant speed in a straight line (Newton's First Law), and then by the statement that the state of motion changes in a predictable manner when "forces" are acting on the body (Newton's Second Law). If one knew the force and mass of the body, the Second Law would determine the instantaneous acceleration of the mass, a, via the relation $F/m = a$. What it meant to speak of the *instantaneous* rate of change of any quantity wasn't (and isn't) obvious, but Newton cleared this up by means of a mathematical innovation, the invention of calculus. From this point forward, mechanics came to mean the study of the motion of masses under the influence of forces described by elaborations of Newton's Second Law, which could now be written as a "differential equation" using calculus.

For this law to be useful, scientists would need to have a mathematical representation of the forces in nature, the F on the left-hand side of $F = ma$. The forces of nature cannot be deduced; they can only be hypothesized (okay, guessed) and tested for whether their consequences make sense and agree with experimental measurements. No amount of mathematical legerdemain can get around that. Newton's Second Law was an empty tautology unless one had an independent mathematical expression for the forces that mattered in a given situation.

Newton gained eternal fame by divining the big one, the one we all know from infancy: the force of gravity. His universal law of gravitation stated that two masses are attracted to each other along the line between their centers, and the strength of that attraction is proportional to the product of their masses and inversely proportional to the square of the distance between them. Of course this attraction is very weak between normal-size masses like two people, but between the earth and a person or the earth and the sun it is a big deal. From this law of gravitation and his Second Law, Newton was able to calculate all kinds of solid-body motion: the orbits of the planets in the solar system, the relation between the moon and tides, the trajectories of cannonballs. Thus Newton had published the first major section of the "book of Nature," which was, Galileo famously declared, "written in the language of mathematics."

Along with the stunning mathematical insights of Newton and their vast practical applications came an ontology, a view of what the fundamental categories of nature were, and how events in the world were related. As Einstein put it in his autobiographical notes, "In the beginning—if such a thing existed—God created Newton's laws of motion together with the necessary masses and forces. That is all. Anything further is the result of suitable mathematical methods through deduction. What the nineteenth century achieved on this basis . . . must arouse the admiration of any receptive man . . . we must not therefore be surprised that . . . all the physicists of the last century saw in classical mechanics a firm and empirical basis for all . . . of natural science."

At the core of the Newtonian view of nature was the concept of rigid determinism, majestically expressed by the Marquis de Laplace:

> We may regard the present state of the universe as the effect of its past and the cause of its future. An intellect which at any given moment knew all of the forces that animate nature and the mutual positions of the beings that compose it, if this intellect were vast enough to submit these data to analysis, could condense into a single formula the movements of the greatest bodies of the universe and that of the lightest atom; for such an intellect nothing could be uncertain and the future just like the past would be present before its eyes.

This Marquis Pierre Simon de Laplace was one of the great masters of classical mechanics in the nineteenth century and became known as the "French Newton." He was willing to literally put his neck on the line for his natural philosophy. When he presented his five-volume study of celestial mechanics to Napoleon, he was greeted with the intimidating question: "Monsieur Laplace, they tell me you have written this large book on the system of the universe and have never even mentioned its Creator." Laplace, normally quite politic in his dealings with influential men, in this case drew himself up and replied bluntly, "I have no need of that hypothesis."

While the relation between mass and the force of gravity was the only fundamental law discovered by Newton, he and other physicists knew that there must be other forces with associated laws, for example, the pressure exerted by a gas (pressure is force per unit area), which surely must have a microscopic origin. Near the end of the eighteenth century Charles Augustin de Coulomb, using a sensitive instrument known as the torsion balance, definitively measured another type of force, also of invisible origin: the electrical force. Coulomb and others determined that, in addition to mass, there is another important property of matter, electrical charge, and that two charged bodies exert forces on each other in a manner similar to the way gravity works in Newton's Second Law, that is, proportional to the product of their charges and inversely proportional to the square of the distance between them. However, there is a major difference between this electrostatic force and gravity; charges come in two types, positive and negative. Opposite charges attract each other, but like charges repel. Matter is usually electrically neutral (that is, made up of an equal number of positive and negative charges) or very nearly neutral, so two chunks of matter don't usually exert much long-range electrical force on each another. Therefore, despite the fact that the electrical force is much stronger than the gravitational force (when appropriately compared), it doesn't have the same kind of macroscopic effects as gravity.

Early in the nineteenth century it became clear that the story was even more complicated. Moving charges (i.e., electrical currents) create yet another force, known to the ancients but not understood as related to electricity: magnetism. Primarily through the work of the English experimental physicist Michael Faraday, it became clear that electricity and magnetism were intimately related because, for example, magnetic fields could be used to create electrical currents. Exploiting this principle, discovered in 1831 and now known as Faraday's law, Faraday was able to build the first electrical generator (he had earlier made the first functioning electric motor). Faraday's discovery would lead to an expansion of the classical ontology of physics, because it implied that electrical charges and currents gave rise to unseen electric and magnetic *fields*, which permeated space and were not associated with matter at all but rather represented a *potential* to exert a force on charged matter. These were the "unseen forces" that moved the compass needle, which had fascinated Einstein as a child. Besides masses, forces, and charges, there were now fields as well.

Faraday had risen from the status of a lowly bookbinder's apprentice to become Fullerian Professor of Chemistry at the Royal Institution (during his life he rejected a knighthood and twice declined the presidency of the Royal Society). When asked by the four-time prime minister William Gladstone the value of electricity, he is said to have quipped,[1] "One day sir, you may tax it." He had little formal mathematical education and showed by experiment that his ideas were correct but did not formulate them into a rigorous theory.

That task was left to the Scottish physicist/mathematician James Clerk Maxwell. Maxwell was a deeply religious man, related to minor nobility, who showed an Einsteinian fascination with natural phenomena from a young age. As early as age three he would wander around the family estate asking how things worked, or as he put it, "What's the go o' that?" He is widely regarded as the third-greatest physicist of all time, after Newton and Einstein, although he is surely much less known to the public. He wrote his first important scientific paper at the age of sixteen and attended Cambridge University, where he excelled and became a Fellow shortly after graduation. One of his contemporaries wrote of him, "He was the one acknowledged genius . . . it was certain that he would be one of that small but sacred band to whom it would be given to enlarge the bounds of human knowledge." At the age of twenty-three Maxwell expressed his philosophy of science in terms that prefigure similar sentiments of both Planck and Einstein:

Happy is the man who can recognize in the work of today a connected portion of the work of life, and an embodiment of the work of Eternity. The foundations of his confidence are unchangeable, for he has been made

---

[1] This wonderful incident may well be apocryphal, as there is no contemporaneous account of it.

a partaker of Infinity. He strenuously works out his daily enterprises, because the present is given to him for a possession.

Maxwell had a full beard and a certain reserved presence that was hard to warm up to (very unlike Einstein, the mensch); however, he was a loyal friend and an almost saintly husband—in all, a man of character and integrity. Despite his diffidence, he possessed a rapier-like wit, which he would only occasionally display, as in the following. In his forties, having "retired" to his Scottish country estate for health and personal reasons, he was convinced to return to England to head the new Cavendish Laboratory at Cambridge; he did a superb job and became an important administrative figure in British science. In this capacity he was asked to explain to Queen Victoria the importance of creating a very high vacuum. He described the encounter thus:

> I was sent for to London to be ready to explain to the Queen why Otto von Guerike devoted himself to the discovery of nothing, and to show her the two hemispheres in which he kept it . . . and how after 200 years W. Crookes has come much nearer to nothing and has sealed it up in a glass globe for public inspection. Her majesty however let us off very easily and did not make much ado about nothing, as she had much heavy work cut out for her all the rest of the day.

The young Maxwell came to know the much older Faraday personally as well as through his work and realized that his experimental discoveries, which Faraday had framed qualitatively, could be cast into a set of equations that describe all electromagnetic phenomena in four compact formulas, now universally known as Maxwell's equations. Like Newton's Second Law these are four differential equations, not describing masses and forces but rather electrical fields, magnetic fields, electrical charges and currents. If Maxwell had used only Faraday's law and the previously known laws of electrostatics and magnetism, he would have found similar equations but with a disturbing asymmetry between the role of the electric and magnetic fields. Maxwell decided in 1861 that these two fields were different expressions of the same unified force, and had the brilliant insight to add a new term to one of the equations describing the magnetic field, which had the effect of making the full set of equations perfectly symmetric in regions of space where there were no electrical charges or currents (as in vacuum). Thus he essentially added a major clause to the laws of electromagnetism. The new term gave rise to new effects, called "displacement currents," which were verified experimentally. They also made the equations structurally perfect. Boltzmann, quoting Goethe, said of Maxwell's equations, "was it God that wrote those lines?"

Having added his new contribution to the electromagnetic laws, Maxwell made a historic discovery: electric and magnetic fields could propagate through the vac-

FIGURE 4.1. James Clerk Maxwell at roughly the age at which he proposed the fundamental laws of classical electromagnetism. Courtesy of the Master and Fellows of Trinity College Cambridge.

uum in the form of a wave that carried energy and could exert both electric and magnetic forces. In physics the term *wave* refers to a disturbance in a medium (e.g., water or air) that oscillates in time and typically is extended, at any single instant, over a large region of space. In this case the strength of the disturbance is measured by the strength of the electric field, so that if an electric charge sat at one point in space the electric field would push the charge alternately up and then down, like a rubber ball bobbing on surface waves propagating through water. Moreover, if you moved along with the wave, like a surfer, the field would always push you in one direction, just as the surfboard stays at the leading edge of a water wave (for a while).

Maxwell showed that the distance between crests of the electromagnetic waves could be made arbitrarily large or small; that is, any wavelength was possible. Thus he discovered what we now call the electromagnetic spectrum, extending, for example, from radio waves having a wavelength of a meter, to thermal radiation (as we saw earlier) at ten millionths of a meter, visible light at half a millionth of a meter, and on to x-rays at ten billionths of a meter. This was a spectacular finding; but the epiphany, the earthshaking revelation, was the speed of the waves: all of them traveled at the same speed, the speed of light! Suddenly disparate phenomena involving man-made electrical devices, natural electric and magnetic phenomena, color, and vision were unified into one phenomenon, the propagation of electromagnetic waves at 186,000 miles per second.

The beauty and significance of this discovery has awed physicists ever since. One of the greatest modern theoretical physicists, Richard Feynman, wrote of this event: "From the long view of the history of mankind . . . the most significant event of the nineteenth century will be judged as Maxwell's discovery of the laws of electrodynamics. The American Civil War will pale into provincial insignificance in comparison with this important scientific event of the same decade." Maxwell himself, with typical understatement, wrote to a friend in 1865, "I have also a paper afloat, with an electromagnetic theory of light, which, until I am convinced of the contrary, I hold to be great guns."

Maxwell would go on to make other major contributions to physics, specifically with his statistical theory of gases, which will be of great relevance below, but he was not recognized as a transcendent figure during his lifetime. He died of abdominal cancer in 1879 at the age of forty-eight, still at the peak of his scientific powers. While in hindsight we view Maxwell as poorly rewarded in his time for his genius and service to society (he was never knighted, for example), Maxwell did not see it that way. On his deathbed he told his doctor, "I have been thinking how very gently I have always been dealt with. I have never had a violent shove in my life. The only desire which I can have is, like David, to serve my generation by the will of God and then fall asleep."

Maxwell's achievement particularly captivated Einstein. Maxwell, Faraday, and Newton were the three physicists whose picture he had on the wall in his study later in life. Of Maxwell he wrote, "[the purely mechanical world picture was upset by] the great revolution forever linked with the names of Faraday, Maxwell and Hertz. The lion's share of this revolution was Maxwell's . . . since Maxwell's time physical reality has been thought of as represented by continuous fields. . . . this change in the conception of reality is the most profound and fruitful that physics has experienced since the time of Newton." Elsewhere he said, "Imagine his feelings when the differential equations he had formulated proved to him that electromagnetic fields spread in the form of . . . waves and with the speed of light"; and, "to few men in the world has such an experience been vouchsafed."

Maxwell had completed the second pillar of classical physics, what we now call classical electrodynamics, to go along with the first pillar, classical mechanics. But neither his nor Newton's equations in themselves answered the fundamental question: what is the universe made of? One knew that there were masses and charges and forces and fields, but what were the building blocks of the everyday world? The enormous challenge was to extend these physical laws down to this conjectured "atomic" scale. Were there new, microscale forces not detectable at everyday dimensions? Did Newton's and Maxwell's laws still hold there? Were atoms little billiard balls with mass and electrical charge obeying classical mechanics and electrodynamics? Were there atoms at all, or were they just "theoretical constructs," as many physicists and chemists maintained until the end of the nineteenth century?

At the time of Maxwell there was no way to probe the internal structure of atoms or molecules directly. As Maxwell put it, "No one has ever seen or handled a single molecule. Molecular science therefore . . . cannot be subjected to direct experiment." However physicists, led by Maxwell and Boltzmann, were beginning to use the atomic concept to explain in great depth the macroscopic behavior of gases. In doing so they were inferring properties of atoms and their interactions. This was the work that Einstein never forgave Herr Weber, his erstwhile mentor, for ignoring. It is here that Einstein first put his shoulder to the wheel.

CHAPTER 5

THE PERFECT INSTRUMENTS OF THE CREATOR

"The Boltzmann is magnificent," Einstein wrote to Maric in September of 1900. "I am firmly convinced that the principles of his theory are right, . . . that in the case of gases we are really dealing with discrete particles of definite finite size which are moving according to certain conditions . . . the hypothetical forces between molecules are not an essential component of the theory, as the whole energy is kinetic. This is a step forward in the dynamical explanation of physical phenomena." Einstein was reading Boltzmann's *Lectures on the Theory of Gases.* The Viennese physicist Ludwig Boltzmann and Maxwell had developed a theory of gases in the 1860s with much the same content, but with the difference that Boltzmann wrote long, difficult-to-decode treatises, while Maxwell's work was much more succinct. Maxwell commented on this drily: "By the study of Boltzmann I have been unable to understand him. He was unable to understand me on account of my shortness, and his length was and is an equal stumbling block to me." Einstein, despite the enthusiasm he expressed to his fiancée in 1900, was later to warn students, "Boltzmann . . . is not easy reading. There are very many great physicists who do not understand it." It is likely that Einstein had no access to Maxwell's work on gases in 1900, and as he did not read English until much later in life, he would not have been able to benefit from it anyway (in contrast, Maxwell's electrodynamics was available to Einstein in German textbooks).

Maxwell beautifully described the scientific advance he had made in atomic theory in an address to the Royal Society in 1873 titled, simply, "Molecules."

> An atom is a body which cannot be cut in two. A molecule is the smallest possible portion of a particular substance. The mind of man has perplexed itself with many hard questions. . . . [Among them] do atoms exist, or is matter infinitely divisible? . . .
>
> According to Democritus and the atomic school, we must answer in the negative. After a certain number of sub-divisions, [a piece of matter] would be divided into a number of parts each of which is incapable of further sub-division. We should thus, in imagination, arrive at the atom, which, as its name literally signifies, cannot be cut in two. This is the atomic doctrine of Democritus, Epicurus, and Lucretius, and, I may add, of your lecturer.

Maxwell goes on to describe how chemists had already learned that the smallest amount of water is a molecule made up of two "molecules" of hydrogen and one "molecule" of oxygen (here he has decided, somewhat confusingly, to use molecule to refer to both atoms and molecules). Then he arrives at his current research.

> Our business this evening is to describe some researches in molecular science, and in particular to place before you any definite information which has been obtained respecting the molecules themselves. The old atomic theory, as described by Lucretius and revived in modern times, asserts that the molecules of all bodies are in motion, even when the body itself appears to be at rest. . . . In liquids and gases, . . . the molecules are not confined within any definite limits, but work their way through the whole mass, even when that mass is not disturbed by any visible motion. . . . Now the recent progress of molecular science began with the study of the mechanical effect of the impact of these moving molecules when they strike against any solid body. Of course these flying molecules must beat against whatever is placed among them, and the constant succession of these strokes is, according to our theory, the sole cause of what is called the pressure of air and other gases.

This simple picture, that gas pressure arises from the collisions of enormous numbers of molecules with the walls of the container, along with simple ideas of classical mechanics, allows Maxwell to derive Boyle's law, that the pressure of the gas is proportional to its density. It also allows him to understand the observation that the ratio of volumes of any two gases depends only on the ratio of temperatures of the gases. The relation of temperature to volume of a gas is critical: in this view absolute temperature (what we now call the kelvin scale) is related to molecular motion and is proportional to the average of the square of the molecular velocity in a gas. Since the energy of motion for any mass, called kinetic energy, is just one-half its mass times the square of the velocity, this also means that for a gas its energy is just proportional to temperature. As Einstein had noted in his letter to Maric, in the Maxwell-Boltzmann theory, the entire energy of a gas is the kinetic energy of moving molecules. This principle of the Maxwell-Boltzmann theory, that the energy of each molecule is proportional to the temperature, applies even in the solid state, in which the molecules vibrate back and forth around fixed positions instead of moving freely throughout the substance. This property of the theory would perplex Einstein later, when he was trying to make sense of Planck's radiation law.

"The most important consequence which flows from [our theory]," Maxwell continues, "is that a cubic centimetre of every gas at standard temperature and pressure contains the same number of molecules." This fact about gases was conjectured by the Italian scientist Amadeo Avogadro in 1811. In 1865 Josef Loschmidt, a professor in Vienna and later a colleague of Boltzmann, had estimated this actual number, which is very large: 2.6×10^{19}, or roughly five billion *squared*. (This "Loschmidt number" is closely related to Avogadro's number, which is the number of molecules in a mole of any gas—both Einstein and Planck were very interested in accurately determining these numbers). With all this information about gas properties, it was possible for Maxwell to determine the average velocity of a molecule in air. He found it to be roughly one thousand miles per hour. He described the implications most picturesquely:

If all these molecules were flying in the same direction, they would constitute a wind blowing at the rate of seventeen miles a minute, and the
only wind which approaches this velocity is that which proceeds from the
mouth of a cannon. How, then, are you and I able to stand here? Only
because the molecules happen to be flying in different directions, so that
those which strike against our backs enable us to support the storm which
is beating against our faces. Indeed, if this molecular bombardment were
to cease, even for an instant, our veins would swell, our breath would leave
us, and we should, literally, expire. . . . If we wish to form a mental representation of what is going on among the molecules in calm air, we cannot
do better than observe a swarm of bees, when every individual bee is flying
furiously, first in one direction, and then in another, while the swarm as a
whole . . . remains at rest.

Maxwell goes on to describe how his own experiments and others
have determined that the molecules in a gas are continually colliding
with one another, moving only about ten times their diameter before
changing direction again through a collision, leading to a kind of random motion called diffusion. Because of this constant changing of
direction, the actual distance moved from the starting point during a
given time is much less than if the molecule were moving in a straight
line. This explained why, when Maxwell took the lid off a vial of ammonia in the lecture, its characteristic odor was not immediately detected in the far reaches of the lecture hall. The same kind of diffusion
occurs in liquids such as water, but much more slowly. Maxwell then
throws off a poetic but profound comment: "Lucretius . . . tells us to
look at a sunbeam shining through a darkened room . . . and to observe
the motes which chase each other in all directions. . . . This motion of
the visible motes . . . is but a result of the far more complicated motion
of the invisible atoms which knock the motes about." Exactly this process occurs to small particles suspended in a liquid but visible under
a microscope, so-called Brownian motion. In one of his four master-
pieces of 1905 Einstein would actually take the suggestion of Lucretius
and Maxwell seriously and, by careful analysis, turn this into a precise method for determining Avogadro's number! Experiments by the

French physicist Jean Perrin would confirm Einstein's predictions and determine that number very precisely; as a result Perrin received the Nobel Prize for Physics in 1926, long after his work had permanently put to rest doubts about the existence of atoms.

The kind of complex, essentially random motion characteristic of gas molecules gave rise to a new way of doing physics, described by Maxwell in the same lecture. "The modern atomists have therefore adopted a method which I believe new in the department of mathematical physics, though it has long been in use in the section of statistics." Thus was born the discipline of *statistical mechanics*. Maxwell could only assume that the invisible molecules obeyed Newtonian mechanics; he had no reason to doubt this. But in describing what would happen in a gas, he realized that one must inevitably encounter the weak point in Laplace's grandiose dictum. Laplace had imagined an intellect that "at a certain moment would know all forces that set nature in motion, and all positions of all items of which nature is composed." Maxwell realized that getting all the necessary information and using it to predict the future was an absurd proposition. "The equations of dynamics completely express the laws of the historical [Laplacian] method as applied to matter, but the application of these equations implies a perfect knowledge of all the data . . . but the smallest portion of matter which we can subject to experiment consists of millions of molecules, not one of which becomes individually sensible to us . . . so that we are obliged to abandon the historical method and to adopt the statistical method of dealing with a large group of molecules." Maxwell's point is that for all practical purposes one doesn't want to know what each molecule is doing anyway; for example, to find the pressure exerted by a gas one needs only to know the *average* number of molecules hitting the wall of a container per second, and how much momentum (mass times velocity) they transfer to the wall.

This was the key insight of Maxwell and Boltzmann: to predict the physical properties of a large aggregation of molecules, one needed only to find their average behavior, assuming they were behaving as randomly as allowed by the laws of physics. Calculating these properties was relatively easy for a gas, where most of the time the molecules are not in close contact; for liquids and solids it was much harder and in certain cases still challenges the physicists of the twenty-first century. Tied up with this insight was a new understanding of the laws of thermodynamics. The First Law says that heat is a form of energy, and that the total energy (heat plus mechanical) always stays the same (is "conserved") even when one form is being changed into the other. For example, when a car is moving at 60 miles per hour, it has a lot of mechanical energy, specifically kinetic energy, $\frac{1}{2}mv^2$, where m is the mass of the car and v is its speed (60 mph in this case). When you slam on the brakes, that kinetic energy doesn't disappear; it is turned into heat in your brakes and tires, due to friction. From the point of view of statistical mechanics, that heat is just mechanical energy transmitted to the molecules of the road and tires, distributed in some complicated and apparently random manner among them. So heat is just random, microscopic mechanical energy, stored in various forms in the atoms and molecules of gases, liquids, and solids.

This view sheds light on the Second Law, which states that disorder always increases and is measured by a quantity called entropy. This law now can be interpreted as saying that in any process where something changes (e.g., the car coming to a stop), you can never perfectly "reorganize" all the energy that goes into the random motion of molecules. It is always too hard to retrieve all of it in a useful form. Before the car stopped, all its molecules (in addition to some random motion due to its non-zero temperature) were moving together in the same direction at 60 mph, providing a kinetic energy that could be used to do useful work, such as dragging a heavy object against friction. As the car stops, that energy is transformed into the less usable form of heat. It is not that we can't turn heat back into usable energy (e.g., use it to get the car moving again); it is just that we can't do it perfectly. We could run some water over the hot brake discs of our stopped car, which could generate steam, which could turn a turbine, and, presto, we would get back some useful mechanical energy. This of course is not the best-designed heat engine one could imagine. But the Second Law says that no matter how carefully or cleverly you design an engine to turn heat into useful mechanical energy, you will always find that you have to put more heat energy in than you get back.

To make this all precise and tractable in a mathematical theory, the German physicist Rudolph Clausius, while a professor at our familiar Zurich Poly in 1865, introduced the notion of entropy, which is a measure of how much the microscopic disorder increases in every process involving heat exchange. The word *entropy* was chosen from the Greek word for "transformation," and indeed Clausius was guided by just the picture we have been painting: heat is the internal energy of atoms or molecules, which can be partially but never fully transformed to usable energy. Now, with their new statistical mechanics, Maxwell and Boltzmann were trying to make this idea of the internal energy of a trillion trillion rocking and rolling molecules precise, and in so doing come to understand entropy and the laws of thermodynamics on the basis of atomic theory. This program was so controversial that even by the end of the century, thirty years later, Planck, the thermodynamicist par excellence, was reluctant to adopt it. It was only his quantum conundrum that forced him to overcome his scruples, as we will see.

The key point is that the statistical mechanics of Maxwell and Boltzmann was still *Newtonian mechanics*, just applied to a system so complicated that one imagines it behaving like a massive game of chance, in which each molecular collision with a wall or with another molecule is like a coin being tossed (heads you go to the right, tails you go to the left). The *worldview* is the same as that of Newton and Laplace; only the method is different. Maxwell, had he lived another two decades, might have begun to recognize the leaks springing in this optimistic vessel, since the basic inconsistency in this view appeared at the intersection of his two great inventions, the theory of electromagnetic radiation and the statistical theory of matter. However, that was not to be; he would pass away a mere five years after his spectacular lecture on molecular science, having spent those final years occupied by his administrative duties. At the end of that same lecture, having anticipated the next twenty-five years of physical theory, the devout Maxwell makes one of the great historical appeals for intelligent design:

Natural causes, as we know, are at work, which tend to modify, if they do not at length destroy, all the arrangements and dimensions of the earth and the whole solar system. But . . . the molecules out of which these systems are built . . . remain unbroken and unworn.

They continue this day as they were created, perfect in number and measure and weight, and from the ineffaceable characters impressed on them we may learn that those aspirations after accuracy in measurement, truth in statement, and justice in action, which we reckon among our noblest attributes as men, are ours because they are essential constituents of the image of Him Who in the beginning created, not only the heaven and the earth, but the materials of which heaven and earth consist.

The next century would demonstrate in many ways, culminating in the awesome demonstration of August 1945, that atoms are not as indestructible as Maxwell had supposed. And Einstein would be the first to understand, through his most famous equation, $E = mc^2$, just how much energy would be released when the perfect instruments of the Creator were disassembled.

CHAPTER 6

MORE HEAT THAN LIGHT

"I have again made the acquaintance of a sorry example of that species—one of the leading physicists of Germany. To two pertinent objections which I raised about one of his theories and which demonstrate a direct defect in his conclusions, he responds by pointing out that another (infallible) colleague of his shares his opinion. I will shortly give that man a kick up the backside with a hefty publication. Authority befuddled is the greatest enemy of truth."

Such was the feisty mood of Einstein as he wrote in July of 1901 to an old friend, Jost Winteler. The object of his ire was Paul Drude, theorist and chief editor of *Annalen der Physik*, the most prestigious physics journal in the world at that time. Drude himself was the author of a well-respected text on optics and Maxwell's equations (in fact it was Drude who introduced the universal symbol c for the speed of light in vacuum). The "infallible" colleague mentioned by Einstein was none other than Ludwig Boltzmann. Einstein, characteristically, seemed oblivious to the potential consequences of offending such prominent scientists, one of whom was editor of the journal to which he would submit all his original research papers for the next six years.

Einstein wrote those lines from the small city of Winterthur, about twenty miles from Zurich, where he had a two-month position teaching physics and mathematics at the Technical College while the regular instructor was performing his military service. The teaching load was quite heavy, thirty hours a week, but, undeterred, he reassured Mileva that "the Valiant Swabian is not afraid." In fact he found that he enjoyed the teaching much more than he had expected, and despite the busy schedule he managed time to study research questions, such as

Drude's new "electron theory of metals." It was only four years earlier, in 1897, that the English physicist J. J. Thomson had confirmed the existence of electrons, negatively charged particles much lighter than the hydrogen atom itself, and he hypothesized that electrons were constituents of atoms. By 1899 Thomson had shown that electrons could be pulled off atoms (a process we now call "ionization") and hence that the atom was in this sense divisible. This represented the first crack to appear in the indestructible atoms of Maxwell.

Drude's theory was based on a guess about the atomic properties of metals. He hypothesized (correctly) that, in a metal such as copper, one of the electrons in each atom was free to move. While the atoms themselves remained fixed in a solid regular crystalline array, these free electrons formed a gas of charged particles that could move easily through the solid, allowing it to conduct electricity and heat efficiently. Drude's hypothesis was that many of the important properties of metals arose from this gas of electrons, and he thus could use the kinetic theory of Maxwell and Boltzmann to calculate those properties. This was an important step forward, and some of Drude's conclusions were based on such general considerations that they remain true and are used in our modern (quantum) theory of metals. Other conclusions he drew from his theory relied on Newtonian mechanics and are now known to be false. Einstein's letter to Drude, pointing out his "errors," and Drude's reply are lost, so nothing is known about the validity of Einstein's objections. What is known is that around this time Einstein began his own reworking of the basic principles of statistical mechanics.

Einstein's very first published work on atomic theory (the one he sent along with his job inquiries) was based on a naive hypothesis about molecular forces: that they behaved similarly to gravity in that they depended only on the distance between molecules, and on the type of molecules involved. He wrote about this work to his former classmate Grossmann in April of 1901: "As for science, I have a few splendid ideas. . . . I am now convinced that my theory of atomic attraction forces can also be extended to gases. . . . That will also bring the problem of the inner kinship between molecular forces and Newtonian action-at-a-distance forces much nearer to its solution." Einstein's simple attraction hypothesis was wrong, and despite his initial enthusiasm he abandoned it after using it in two articles for *Annalen der Physik*, which he later referred to as "my worthless first two papers." These works did emphasize that at the time there was no real understanding among physicists about the origin of molecular forces. After the modern atomic theory was established, it became clear that all the atomic forces important for chemistry or solid-state physics ultimately arise from electromagnetic forces; there *are* no special new molecular forces.[1] However, how atoms *behave* under the influence of these forces is quite different from what was expected, because they obey a new mechanics (quantum mechanics) and not the classical mechanics of Newton. (In addition, there are new forces within the atomic nucleus, hinted at by the phenomenon of radioactivity, which had just been discovered, but these forces are generally not important for chemistry or solid-state physics.) Einstein praised Boltzmann's statistical theory of gases precisely because it *didn't* rely much on the unknown molecular forces, and after his first immature efforts he decided to pursue the path of statistical mechanics into the atomic realm.

At the end of 1901 Einstein received "A letter from Marcelius [Marcel Grossmann] . . . a very kind letter" telling him that the patent office position would soon be advertised and that he would definitely get it. "In two months time we would then find ourselves in splendid circumstances and our struggle would be over," he wrote to Maric, but, he hastened to add, their bohemian lifestyle would not change "We shall remain students [horribile dictu] as long as we live, and not give a damn about the world." One more professional disappointment remained. In November 1901 Einstein submitted a PhD thesis on the kinetic theory of gases to Professor Alfred Kleiner at the University of Zurich (the Poly was not yet able to grant PhD degrees, although with Weber in charge it is not likely that Einstein would have tried

---

[1] There are forces inside the atomic nucleus that were unknown at that time, now called, unimaginatively, the "strong" and "weak" nuclear forces. At that time the existence of the nucleus was itself unknown.

that route anyway). Only indirect information survives about what transpired, but Einstein withdrew the thesis early in 1902, apparently at Kleiner's suggestion, "out of consideration for [Kleiner's] colleague Ludwig Boltzmann," who, despite Einstein's admiration for his work, had been "sharply criticized" on certain points.

By February of 1902 Einstein had relocated to Bern, a picturesque Swiss city on the river Aare. His job at the patent office would not start for five months, and he remained without visible means of support, so he literally hung out his shingle.

> Private Lessons in
> MATHEMATICS and PHYSICS
> for students and pupils
> given most thoroughly by
> ALBERT EINSTEIN, holder of the fed.
> polyt. teacher's diploma
> GERECHTIGKEITGASSE 32, 1st floor
> Trial Lessons Free

"The situation with the private lessons isn't bad at all. I have already found two gentlemen, an engineer & an architect & more in prospect," Einstein wrote to Maric shortly after arrival. His letter was apparently a bit *too* cheerful, as he soon received a reply from his fiancée, which is lost, but the content of which is clear from Einstein's rapid follow-up: "It is true . . . that it is very nice here. But I would rather be with you in some backwater than without you in Bern." Actually, although Einstein painted a rosy picture of his life in Bern, an old family friend who visited him there described his condition as "testifying to great poverty . . . [living in] a small, poorly furnished room." Strikingly, Einstein never complains in his letters about material conditions, making only occasional humorous allusions to this "annoying business of starving."

In Bern Einstein quickly gathered around him a lively circle of friends with shared intellectual interests, several of whom would become lifelong companions. By the end of June 1902 he had taken up his post at the patent office as an expert third class (the lowest rank), and his immediate financial woes were ended. In October of that year he obtained grudging permission from his parents (at his father's deathbed) to marry Mileva, and on January 6, 1903, with no family present, only two friends, the couple were married with complete lack of ceremony at the Bern registry office. Typically, Einstein had trouble getting into their new apartment that night, as he had forgotten the key. Mileva had gone through many tribulations to get to this point, and it seems that after failing twice to get her teacher's degree from the Poly, she had now given up her own scientific ambitions. Einstein reports to his friend Michele Besso that she takes good care of him and he "leads a very pleasant, cozy life" with her; in the same letter he tells Besso that he has just sent off his second paper on statistical mechanics, pronouncing the paper "perfectly clear and simple, so that I am quite satisfied with it."

The main point of Einstein's first two papers on statistical mechanics was to frame all the statistical relations that ultimately underlie the First and Second Laws of thermodynamics in a very general way, no longer referring specifically to gases and collisions in gases, as was done by Maxwell and Boltzmann. Thermodynamics is supposed to apply to everything that stores energy and can absorb or give off heat, which is essentially everything: liquids, solids, machines, . . . you name it. The Second Law of thermodynamics implies that no engine can change heat into useful work with perfect efficiency. Stated more picturesquely, it is impossible to make a perpetual-motion machine, a machine that, once it gets started, will go forever in a repeated cycle without needing fuel. Einstein, in his new job at the patent office, was regularly coming across proposed "inventions" that, upon closer inspection, were physically impossible because they violated this principle. The generality of the laws of thermodynamics must have been very much on his mind.

And so he wrote two papers that assume almost nothing about the nature of molecular forces, or the macroscale nature (e.g., gas, solid, etc.) of the thermodynamic system being considered, and that lead to several equivalent forms of the Second Law. The papers make only one assumption, an assumption so subtle that it had been the cause of debate since the time of Maxwell. Einstein appears to have been unaware of this raging debate and does not emphasize this assumption (to be discussed later) or comment on it in any detail. However, the mathematical results of these papers and the formal framework he introduces are quite important, and would alone have made his name known a century later, except for some bad luck.

Independently, and earlier, Josiah Willard Gibbs at Yale University had established exactly the same principles ("the resemblance is downright startling," Max Born later commented) and applied them very powerfully to chemical problems. Gibbs, the son of an eminent theologian and scion of an old New England family, received in 1863 the first PhD in engineering granted in the United States. He briefly studied in Europe and became acquainted with the nascent German school of thermodynamics, begun with Clausius and continuing in Einstein's day with Planck. Very reminiscent of Maxwell in his breadth of interests, Gibbs would make enormous contributions as a physicist, chemist, and mathematician until his death in 1903; he is arguably the greatest American-born scientist of all time. In fact Maxwell himself was so impressed with a clever geometric method devised by Gibbs to determine chemical stability that he made a plaster model illustrating the idea with his own hands and sent it to Gibbs.

Gibbs introduced the concept of "free energy," which dominates modern statistical mechanics and is often denoted by the symbol G in honor of Gibbs; this is just one of *twelve* important scientific contributions bearing his name to this day. His work was initially slow in becoming known in Europe, but just as Einstein was beginning his own statistical studies, Gibbs's monumental treatise, *Elementary Principles of Statistical Mechanics*, was published, and he was awarded the Copley Medal of the Royal Society of London. (Before the Nobel prizes, which were first awarded in 1901, this was the most prestigious international science award of the day.)

Gibbs's contributions predated and overwhelmed those of Einstein, and Einstein would later comment in print that had he known of Gibbs's work earlier, he would "not have published those papers at all, but confined myself to a the treatment of some few points [that were distinct]." Einstein's admiration for Gibbs remained so great throughout his life that when, a year before his death in 1955, he was asked who were the most powerful thinkers he had known, he replied: "[Hendrik] Lorentz,[2] I never met Willard Gibbs; perhaps, had I done so, I might have placed him besides Lorentz." So already, before his twenty-fifth birthday, Einstein had established himself as a deep thinker, on par with the great leaders of his era; unfortunately no scientist of any influence seems to have noticed this at the time. Moreover he was not advancing his career aspirations by telling the leading physicists in Germany of the errors they had made in their earlier work. He would need to devise new theories, which made specific experimental predictions, to get the world's attention. These would not be long in coming, and when they did come, the free-spirited bohemian outsider would soar above even the great men—Maxwell, Gibbs, and Lorentz—whom he so admired.

---

[2] Hendrik Lorentz, a Dutch physicist, was widely regarded as the most eminent theorist of the generation preceding Einstein's; he will play an important role in our story below.

CHAPTER 7

DIFFICULT COUNTING

The tomb of Ludwig Boltzmann in Vienna is engraved with a very short and simple-looking equation, which, ironically, he never wrote down during his lifetime:

$$S = k \log W.$$

S is the universal symbol for entropy, k is a fundamental constant of nature known as Boltzmann's constant, and $\log W$ is the logarithm[1] of a number, W, relating to the physical system of interest, a number that Boltzmann called the number of "complexions." (The number W can be so devilishly hard to calculate for many physical systems of interest that the greatest mathematical physicists of the twenty-first century, and the most powerful computers as well, are helpless to determine it.) This equation was the lever for setting the quantum revolution in motion. It would form the basis for Planck's derivation of his radiation law and for Einstein's first insights into the quantum nature of light.

Clausius had introduced the concept of entropy to explain heat flow, but he had no idea how a physicist could calculate this quantity from any fundamental mechanical theory (presumably an atomic theory). It was the statistical mechanics of Boltzmann, Maxwell, Gibbs (and Einstein) that gave the recipe. The recipe is deceptively simple sounding.

---

[1] Experts will know that in this equation the base of the logarithm is not the usual base 10 version, but is what is called the natural logarithm. The difference is not essential for understanding the meaning of entropy.

FIGURE 7.1. The epitaph on the tomb of Ludwig Boltzmann, S = k log W, expressing the fundamental equation of entropy, which he had discovered. Image courtesy Daderot.

It has two parts: (1) Whatever can happen, will happen. (2) No atom (or molecule) is special.

Imagine shooting gas molecules one by one through a small hole into a box where they bounce around. Mentally divide the box into two equal parts with an imaginary partition. Any part of the box is equally accessible to the molecules (whatever can happen, will), and there is no reason for any molecule to be at one place at a given time versus another (no molecule is special). Suppose for the moment you can actually see the molecules directly. With one molecule in the box you look in periodically, and you find it roughly half the time on the left side of the box and half the time on the right side. Now add a second molecule, wait a bit, and start looking again. Roughly one-fourth of the time both molecules are on the left, one-fourth of the time both are on the right, and half the time one is on the left and one is on the right. Why is the last case more likely than the first two? Because there are two ways that you can get the last case (molecule 1 on right, molecule 2 on left; molecule 2 on right, molecule 1 on left) but only one way you can get the first two cases. This is just like tossing two coins and finding that one heads and one tails happens roughly twice as often as heads-heads and tails-tails. We can now get fancy and define three "states" for the two-molecule gas: in state one, both are on the left; state two, both on the right; and state three, one on the left and one on the right. For the first two states Boltzmann's $W=1$ (there is only one way to get these states), but for the third state[2] $W=2$. The entropy of this third state is then larger than that of the other two states, according to Boltzmann's formula (we need not delve into the mysterious properties of the logarithm function to reach this conclusion). In actuality a physicist would specify the states in more detail than just which half of the box the molecules are in, but the underlying concept and method is exactly the same.

Now imagine we have a few trillion trillion molecules in the box (as indeed we usually do). There is still only one way to have all of them on one side; however, W, the number of ways of having about half on the right side and half on the left side, is unspeakably large. We literally have no words, no analogies, for numbers of this magnitude. As a feeble attempt, imagine the following: take all the atoms in the universe and, in one second, clone each of them, so as to create a second "universe." Now repeat this every second, creating 4, 8, 16, and so on "universes." Do this every second for the *entire age* of our current universe. Add up all of the atoms in all these universes and you will arrive at a number that is incredibly big, all right, *but still* this number is negligibly small compared with the number of states of high entropy of *one*

---

[2] In chapters 24–25 we will learn that under certain circumstances the method for counting the states of quantum gases can differ from this classical reasoning. However, Boltzmann's equation for entropy still holds, just with a different counting method for W.

liter of gas. These high entropy states, in which the gas molecules are roughly equally distributed in each half, have (not surprisingly) enormously higher entropy than the states in which the molecules are all or mostly on one side of the box.

Suppose that we go to a lot of trouble and evacuate the gas from the box, put an airtight partition in the middle, and put the gas back in on the left side, so that we set the system up in this very improbable (low entropy) state. From Maxwell we know that the gas molecules are flying around at 1,000 mph, colliding with one another and the walls and thus creating pressure on the walls. If we then remove the partition, the gas will rapidly fill the entire box again, approximately equally in each half. The entropy of the system will have increased. After the molecules spread themselves out roughly equally in the box, the molecules will still be colliding and moving around, but on average there will be roughly equal numbers on each side of the box. Intuitively this situation is the most disordered state (you don't need to make special efforts to achieve it), and according to Boltzmann's principle, this is the state of maximum entropy. This is the atomic explanation for the Second Law of thermodynamics, that entropy always increases or stays the same. Whenever we try to generate useful work from heat, we are essentially trying to create order out of this molecular chaos, and we are fighting against the laws of probability.

Consider further the previous example, where we have opened a partition and let the gas fill the entire box instead of just half the box. Now it *is* possible that if we wait long enough, all the collisions could work out just right, and all the gas molecules could reconvene on the left side. Is it worth waiting for this to occur? Not really. One can easily calculate that if all the states are equally likely and if we have only *forty* gas molecules in the box, it would take about the age of the universe for this to happen. With a trillion trillion molecules in the box? As they say in New York: fuggedaboutit.

This was the subtle point that Einstein missed in trying to prove that the Second Law of thermodynamics, the increase of entropy, is an absolute law. It isn't absolute; entropy is *allowed* to decrease. Just don't bet on it.

In fact, this overwhelmingly probable increase of entropy is how we determine the direction of time. Imagine that the gas in our box is colored and hence visible as it expands; if we saw a movie of the gas contracting back into half of the box, we would immediately assume that the movie was being run backward. Because the arrow of time is so fundamental, it was natural for physicists to assume that the increase of entropy was an absolute law of nature and not just a very, very, very, very, very . . . likely occurrence. When the young Einstein made this mistake, he was in good company; Boltzmann also got this wrong until his critics pointed it out to him. However, the canny Scot, Maxwell, was not fooled, and described the situation with colorful imagery: "the Second Law of Thermodynamics has the same degree of truth as the statement that if you throw a tumblerful of water into the sea, you cannot get the same tumblerful of water out again."

Maxwell invented an imaginary creature, dubbed "Maxwell's demon," to illustrate this point further. His demons were "lively beings incapable of doing work" (i.e., of adding energy to the gas). He imagined these miniature sprites hovering in the gas, which is uniformly distributed in the box, but now a partition is added to the box, in which the demon has cleverly fashioned a frictionless trapdoor. Whenever the demon sees a gas molecule coming at him from the right to left with high velocity, he lets it through, then closes the trapdoor before any molecules can escape from the left to right. In this way over time he groups the faster molecules on the left and the slower ones on the right. But for a gas the temperature is proportional to the average energy, so by doing this the demon has heated the left side and cooled the right side, *without putting any energy in*. In other words the demon has created a refrigerator on the right (and a heater on the left), neither of which require any fuel (Einstein definitely would have rejected this patent). And why did Maxwell create his demons? Not as a serious proposal for an invention. Instead his intention was "to show that the Second Law of Thermodynamics has only a statistical certainty." Maxwell's demons, spawned around 1870, did not fancy the trip across the Channel, and so the true meaning of the Second Law was not understood in Europe for several more decades.

While Einstein did not recognize the "demonic" exception to the Second Law when he was reinventing Gibbs's statistical mechanics in 1903–4, he did very much focus on what he called "Boltzmann's principle," the mathematical epitaph, $S = k \log W$, mentioned above. In his third paper on statistical mechanics in 1904, he states, "I derive an expression for the entropy of a system, which is completely analogous to the expression found by Boltzmann for ideal gases [$S = k \log W$] *and assumed by Planck in his theory of radiation*" (italics added).Later in that same paper he explicitly applies his results to Planck's thermal radiation law, although in a manner that doesn't yet refer to the quantum concept. This made Einstein the first physicist to extend the use of Planck's law and to accept that statistical mechanics, which had previously been used only to describe gases, could also explain the properties of electromagnetic radiation. Radiation at this point was conceived to be a purely wave phenomenon, having nothing in common with the aggregate of particles (molecules) that make up a gas. Einstein now analyzed thermal radiation using his statistical methods, and he was beginning to see the problems with Planck's "desperate" solution.

So what *had* Planck actually assumed about radiation, and how had he used Boltzmann's principle to justify the formula that he had initially guessed by fitting the data? Planck had not been as bold as Einstein; he did not apply statistical mechanics to radiation but rather to the matter that exchanged energy with radiation. The Planck radiation law is, strictly speaking, only completely correct for what physicists call a "blackbody." We all learn in school that the color white is a mixture of all colors and that black is the absence of color. A perfectly black body absorbs all light that falls upon it; hence no light of any color is directly reflected from it, and it appears black. In contrast, a surface that looks blue to us absorbs most of the red and yellow light incident on it and reflects the blue to our eye. But does the black object actually emit no light? Well, yes and no. It doesn't emit any *visible* light, but it does send out a lot of electromagnetic radiation; however, as we learned earlier, if the object is at room temperature, the radiation is mainly at infrared wavelengths, which we can't see. As already mentioned, the radiation law is precisely the rule for how much EM radiation of a given wavelength a blackbody emits at a given temperature.

To test this ideal behavior, physicists had to find a perfectly black body, not just for visible radiation but for all possible wavelengths. Unfortunately all real materials reflect EM radiation at *some* wavelengths, so soot, oil, burnt toast, and the other obvious candidates don't actually do the job. So the experimenters came up with a clever idea: instead of using the surface of a material, they would use the inside of a kind of furnace with a small hole. Any radiation that went in through the hole would bounce around, being reflected many times, but eventually it would be absorbed before escaping. Thus any light coming out of the hole must have been emitted from the walls and would be representative of a perfect blackbody.

It was this kind of ideal black box or "radiation cavity" that Planck analyzed between 1895 and 1900. And one of his first ideas was to transfer his ignorance of the blackbody law from radiation to matter. He assumed that the walls of the cavity were made of molecules that would vibrate at a certain frequency in response to the EM radiation that fell upon them. Then by a clever argument he related the density of the energy of EM *radiation* at a given frequency (his goal), to the average energy of the vibrating *molecules* at the same frequency.[3] He thus no longer had to deal with Maxwell's equations describing the electromagnetic waves; he could assume that Newton's laws held for the molecular vibrations, and he could use statistical mechanics. However, instead of doing the obvious thing and calculating the average energy of a molecule from statistical mechanics à la Boltzmann, he chose to find the *entropy* of the molecules.

He did this for a strange historical reason. When he began studying blackbody radiation in 1895, he hoped to find the missing principle that would restore the perfection of the Second Law and make the increase of entropy and hence the arrow of time an absolute principle of nature. While he had been convinced that the equations of matter

---

[3] Planck did not call his vibrating entities molecules but used the term "resonators" instead to emphasize that they were idealized microscopic oscillators and that he was not committing himself to any atomic theory. At this point the composition of the atom, with a compact nucleus and electrons bound to it, was not known, although, as we saw from Maxwell, the concept of atoms and molecules was widely accepted by the leading statistical physicists of the time.

allowed for entropy to decrease (although very rarely!), he hoped that those of Maxwell would prevent this from ever happening. This turned out to be a vain hope, as Boltzmann himself was able to demonstrate to Planck. However, having committed himself to the study of the entropy of radiation, and since the actual radiation law was still not definitively known, Planck continued his investigations. He knew that if he could find the average entropy of thermal radiation, it was related by straightforward mathematical steps to the average energy density, and hence to the correct radiation law.

At first Planck made an incorrect argument, which did not rely on Boltzmann's principle. This led him to what was then called the Planck-Wien[4] radiation law, and the embarrassing retraction in October 1900 after that law was ruled out by the experiments of his friends Rubens and Kurlbaum. At that point, guided by their experimental results and his mathematical intuition, as we saw earlier, he guessed the right form of the radiation law. Now, working *backward* from his apparently correct guess for the energy density of the radiation, he could figure out what the corresponding mathematical expression for the entropy of the radiation *had to be*. So this distinguished physicist was in a position oddly familiar to novice physics students, who might find the correct answer to a problem listed in the solutions at the back of their textbook but can't quite figure out how to get that answer based on the principles they are supposed to have learned.

Faced with this quandary, for the first time in his career Planck resorted to Boltzmann's principle. By accepting and using that principle (the formula $S = k \log W$), he now had an approach to justify his empirical guess from the fundamental laws of statistical physics. What he needed to do was to count the possible states of molecular vibration, W, and show that when plugged into Boltzmann's formula, it gave the answer that he "knew" was correct.

The mathematical problem he faced can be posed as follows. Planck assumed that all the molecules in the walls of the blackbody cavity had

---

[4] Recall that after Planck came up with his new radiation law, which agreed with experiment, his name became attached to the new correct law and was dropped from the older law, now referred to as simply Wien's law.

a fixed total amount of energy, which we can think of as a quantity of liquid, such as ten gallons of milk. For simplicity, imagine that there are one hundred molecules in the walls and that each molecule corresponds to a container that can hold up to the entire ten gallons. The question is how many ways can the ten gallons be shared among the hundred containers? If milk (and energy) are assumed to be continuous, infinitely divisible quantities, then the obvious answer is an infinite number of ways. But this didn't deter Planck. The number of places you can put a gas molecule in a box is also infinite, but Boltzmann had found that his answer for the entropy of a system didn't depend in any important way on how he divided the box into smaller boxes. So Planck essentially put little tick marks on the molecular energy containers, saying, for our imaginary example, that milk could only be distributed one fluid ounce at a time. Now he could go ahead and calculate the finite number of ways the milk could be shared and how that number depended on both the total amount of milk (the energy), the number of containers (molecules), and the size of the tick marks (the minimum "quantum" of energy). He was expecting that, as for Boltzmann's gas calculation, nothing crucial would depend on the size of the tick marks. He was mistaken.

Try as he might, if he let the spacing of the tick marks get smaller and smaller, the calculation yielded the wrong entropy and the wrong radiation law. Finally he was forced to the conclusion that there must be some smallest spacing of the tick marks, that is, that energy could only be distributed among the molecules in some smallest "quantized" unit. Since there was absolutely zero justification for this final hypothesis, it is clear why Planck called it "an act of desperation." To his credit, however, Planck did not shy away from stating clearly his unprecedented conclusion in his famous lecture of December 14, 1900, on the blackbody law:

> We consider, however—*this is the most essential point of the whole calculation*—[the energy] E to be composed of a very definite number of equal parts and use thereto the constant of nature $h = 6.55 \times 10^{-27}$ erg-sec. This constant, multiplied by the frequency v . . . gives us the energy element, ε.

Now we can understand fully this cryptic statement. The "definite number of equal parts" were the "tick marks," that is, the minimum quantum of molecular vibrational energy, ε. Moreover it was clear from other considerations that in order to get the right radiation law, this minimum energy must be proportional to the frequency at which the molecules vibrated; thus he was forced to the conclusion that $\varepsilon = h\nu$. Here the Greek letter ν stands for the vibration frequency, and h (as Planck says) is a new constant of nature, undreamt of in our previous natural philosophies. Finally, because the radiation law was measured experimentally, he could go to the data and quickly figure out the actual value of the constant h (quoted above), which is now known as Planck's constant and is the signature of all things quantum.

Planck later said that the radiation law had to be justified "no matter how high the cost." Although he didn't emphasize it at the time, the cost *was* very high. Planck's little, technical fudge, if taken seriously, said something very, very strange about forces and motion at the atomic scale. It said that the Newtonian picture could not be right. For all intents and purposes, Planck had described molecules as little balls on springs, which stored energy by being compressed, and when the springs vibrated the energy was transferred back and forth between this stored (potential) energy and the kinetic energy of motion of the molecules, but in such a way that the sum of these energies, the total energy, was conserved. This much is standard Newtonian physics.

But in Newtonian physics the initial amount of total energy can vary continuously; all you need to do is compress the spring a little more, and it will have a little more energy. The fact that it can have any amount of energy (between some limits) appears intuitively to be related to the very fact that space is continuous. Nothing in Newtonian physics could explain quantized amounts of energy, the idea that the spring could only be compressed, say, precisely 1 or 2 or 3 or . . . inches but nothing in between. This was like imagining a car that can only go 0, 10, 20, . . . miles per hour and nothing in between. The obvious question is: how does it get from 0 to 10 miles per hour without passing through the intermediate values as it accelerates?

There was nothing innocent about Planck's explanation of the radiation law. If it were the real explanation, it was a time bomb hidden in a thicket of algebra, which would explode with earth-shattering implications. *Atoms and molecules were* not *little Newtonian billiard balls; they obeyed completely different and counterintuitive laws.*

But Planck did not insist that his quantum hypothesis was a statement about the real mechanics of actual molecules. In fact he dropped a small hint in his lecture that perhaps energy is not *really* quantized. He denoted the total energy of his molecules as E and stated, "dividing E by ε we get the number P of energy elements which must be divided over the N resonators [molecules]. *If this ratio is not an integer, we take for P an integer in the neighborhood*" (italics added). But if molecular vibrations were *really* quantized, then E/ε would have to be a whole number! Planck was hedging his bets, signaling that one didn't have to take this crazy energy element too seriously. Planck thought the constant of nature he had discovered, h, was very important, but there is no evidence that he believed his derivation invalidated Newtonian mechanics on the atomic scale.

Why not? Theoretical physics is a tricky business; sometimes one can get the right answer with assumptions that are wrong, or at least with stronger assumptions than one really needs. Perhaps another line of argument would occur to Planck, one that would preserve the welcome constant h but dispense with the uncomfortable assumption of the energy quantum, ε. Perhaps this weird, apparent quantization of energy only involved the interaction of radiation with matter but not mechanics per se. After all, there had been no obvious evidence of Planck's constant in other areas of physics. It could be a new embarrassment if he trumpeted this energy quantum as a breakthrough in atomic physics and it turned out not to be so. No, best to play it safe, thought Planck; no need to cry wolf.

So, remarkably, Planck said nothing more in print for five full years about his great discovery, and the strange assumption buried in his derivation remained almost unnoticed. Except in Bern. There the unknown patent clerk's searching investigations into the foundations of statistical mechanics were placing Planck's Rube Goldberg mechanism on the witness stand and returning a verdict: not innocent.

CHAPTER 8

THOSE FABULOUS MOLECULES

One of the great open questions in the history of science is how Einstein came to the core idea of his paradigm-shifting paper of 1905. No, not his paper on special relativity or his paper proposing the famous equation $E = mc^2$. Einstein was asked over and over again how he had developed the key insights leading to the special and general theories of relativity, and he answered with various charming anecdotes that have become part of his legend. As far as we know, he never went on record as to how he came up with the basic conception for his first paper of the annus mirabilis, a radical alternative to Maxwell's theory of electromagnetic waves, which is the only one of his discoveries that he himself labeled as "revolutionary." He says nothing directly about how he arrived at his first work on quantum theory in either his contemporary correspondence or in the papers preceding it. However, there are a few clues in the historical record, and these suggest that the key insight was his realization that the Planck radiation law was absolutely incompatible with statistical mechanics, at least in the form developed by Maxwell, Boltzmann, and Gibbs. This understanding likely matured during the year 1904 and early in 1905, when he was living a comfortable married life with Mileva and, as he was unknown to the wider physics community, his scientific correspondence was quite thin, leaving few traces of his profound ruminations.

As already mentioned, by 1903 Einstein had settled into his routine in Bern, working six days a week at the patent office, giving private lessons, and nonetheless finding time to pursue research in fundamental physics. Later he would refer to this period as "those happy Bernese years." With his charisma and joie de vivre he had very quickly acquired a group of comrades who would share this idyllic interlude with him. The first of these new companions was a Romanian philosophy student, Maurice Solovine, who showed up at his flat in response to Einstein's earnest advertisement for private physics lessons. A typical Einsteinian episode ensued. Following an enthusiastic invitation to enter his humble abode, Solovine was immediately "struck by the extraordinary brilliance of his large eyes." Two and a half hours passed in a twinkling as the men discussed science and philosophy, and by the next session Einstein, having quite forgotten the original profit motive, declared physics lessons too much of a bother and proposed instead that they should meet freely to discuss ideas of all sorts. Very soon they added to their ranks another young aspiring intellectual, Conrad Habicht, a mathematics student, who had attended the Poly a bit ahead of Einstein and whose acquaintance Einstein had made during his vagabond years after graduation.

Habicht had the most jovial and high-spirited relationship with Einstein of all his peers; their letters to each other are rife with playful sarcasm. Together with Solovine, the two men founded a reading and discussion group, which they satirically dubbed the "Olympia Academy." Habicht graciously allowed Einstein the esteemed position of president, complete with a commemorative (cartoon) bust and a grandiloquent dedication in Latin, celebrating his unerring command of "those fabulous molecules." It was also Habicht who dubbed our Valiant Swabian "Albert Ritter von Steissbein," which loosely translates as "Knight of the Tailbone," presenting him with an engraved tin plate bearing this title. Far from being offended, Einstein and Mileva "laughed so much they thought they would die," and henceforth Albert occasionally signed letters to Habicht with this sobriquet. The heraldic crest above his bust is aptly chosen: a link of sausages, one of the few foodstuffs the Olympians could afford to eat at their august gatherings.

Despite the evident joviality of the meetings, the members, along with occasional guests such as the attentive but silent Mileva, took their studies very seriously, and Einstein acquired many of his lasting

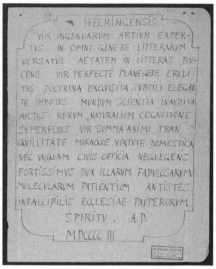

FIGURE 8.1. (a) Hand-drawn cartoon by Maurice Solovine celebrating Einstein as President of the Olympia Academy, with his bust garlanded in hanging sausages. (b) Satirical inscription in Latin that accompanied the cartoon. It translates as "The man of Hechingen, expert in the noble arts, versed in all literary forms – leading the age towards learning, a man perfectly and clearly erudite, imbued with exquisite, subtle and elegant knowledge, steeped in the revolutionary science of the cosmos, bursting with knowledge of natural things, a man with the greatest peace of mind and marvelous family virtue, never shrinking from civic duties, the powerful guide to those fabulous receptive molecules, infallible high priest of the poor in spirit." Courtesy the Albert Einstein Archive.

philosophical views during the two years of meetings. The group would convene at the apartment of one of the members, and over a frugal repast would debate the meaning and merits of the assigned works, which included philosophy (David Hume and John Stuart Mill), history and philosophy of science (Henri Poincaré and Ernst Mach), and occasionally great literature (*Don Quixote* and *Antigone*).

Solovine recalled that these gatherings brimmed over with merriment, although woe to him who would slight the gravity of the occasion. On one memorable night, Solovine, having skipped out on a meeting at the last minute to attend a concert, returned to his apartment to find his bed piled high with furniture and household items, his room enveloped in thick cigar smoke, and a scolding note in Latin pinned to his wall. Einstein remembered these get-togethers with the greatest fondness. In 1953 he wrote to Solovine:

FIGURE 8.2. The principals of the Olympia Academy circa 1903; on the left is Conrad Habicht, center is Maurice Solovine, and on the right is Einstein. ETH-Bibliothek Zurich, Image Archive.

To the immortal Olympia Academy!

In your short active life dear Academy, you took delight, with childlike joy, in all that was clear and intelligent. Your members created you to make fun of the long-established sister academies. How well your mockery hit the mark I have learned to appreciate through long years of careful observation. . . . Even if we have become somewhat decrepit, a glimmer of your bright, vivifying radiance still lights our lonely pilgrimage. . . . To you our fidelity and devotion until the last learned gasp!

A. E.—now corresponding member

Besides Habicht and Solovine, Einstein had another intellectual confidant in Michele Besso. He had met Besso, who was six years older and already a graduate of the Poly, while he was a student in Zurich. Besso, of middle-class Jewish origin, was an engineer, who had applied for and obtained a position in the patent office in Bern at Einstein's suggestion. Einstein had also introduced him to his eventual wife, Anna Winteler, who was a daughter of Einstein's host family in Aarau. Besso and Einstein became lifelong and intimate friends. If Einstein occasionally exhibited the absentmindedness of a starry-eyed dreamer, he was an absolute model of Swiss efficiency compared with the impractical Besso, whose boss had once pronounced him "completely useless and almost unbalanced." Einstein, while granting that in many respects Michele was "an awful schlemiel," nonetheless enjoyed and profited from their exchanges: "[Besso] has an extraordinarily fine mind whose working, though disorderly, I watch with great delight." Later, when reflecting upon his achievements of 1905, he said that he "could not have found a better sounding-board in the whole of Europe." Besso and Einstein walked home together daily during that period, and Einstein shared with him his developing ideas about physics. Although Besso is often mentioned as the first to hear about relativity theory, apparently the main subject of their conversations was something else: Einstein's new hypothesis about the nature of light.

This hypothesis surely would have been presented to the Olympia Academy; however, by the end of 1904, Habicht had obtained a post as a mathematics and physics teacher in Schiers, quite a distance from Bern, and in 1905 Solovine moved from Bern to Paris, where he worked on a journal, the *Revue Philosophique*. Hence the academy was out of session as Einstein was producing his string of hits. In May of 1905 he sent a typical jocular missive to Habicht, whose absence he had felt:

> Dear Habicht, such a solemn air of silence has descended between us that I almost feel as if I am committing a sacrilege when I break it now with some inconsequential babble. . . .
>
> So what are you up to you frozen whale, you smoked, dried, canned piece of soul . . . ? Why have you still not sent me your dissertation? Don't you know that I am one of the $1\frac{1}{2}$ fellows that would read it with interest and pleasure, you wretched man?
>
> I promise you four papers in return, the first . . . deals with radiation and the energy properties of light and is *very revolutionary*. . . . The fourth

paper is still only a rough draft at this point, and is an electrodynamics of moving bodies which employs a modification of the theory of space and time.

This lively discourse indicates Einstein's contemporaneous valuation of his great works of 1905; they were all exciting, and he was proud of them, but one was actually revolutionary, the one on the quantum properties of light. This paper, titled "On a Heuristic Point of View Concerning the Production and Transformation of Light," grew out of his last work on statistical mechanics, in which he had focused for the first time on Planck's theory of radiation, the theory that required the critical but unappreciated introduction of a smallest energy element, $\varepsilon = h\nu$.

Recall that Planck had avoided the question of whether he was allowed to apply statistical mechanics to radiation by relating the radiation energy, which he wanted to calculate, to the average energy of vibrating molecules in the blackbody. But instead of directly calculating this energy, he took the odd detour of calculating the *entropy* of the molecules, and, with full knowledge of the answer he *had* to get in order to agree with experiment, he wrestled the entropy into the correct form, using the outré assumption of a minimal energy element. Given Einstein's focus on statistical mechanics, which, unaware of Gibbs's prior work, he thought he was extending in a novel manner beyond gases, he would naturally focus on the place at which Planck took his detour. Why not just calculate the average energy of a molecule?

In this context, what is needed is the average "thermal energy" of the molecule. If we imagine the molecule as consisting of several atoms connected by chemical bonds, which can vibrate just like macroscopic springs, and if it were possible to direct energy to a particular isolated molecule, then according to the classical view one can tune the energy to any desired value (until it gets so large that the molecule breaks apart). However, Planck had appended the ad hoc restriction that each vibration could only have certain energy values, constrained to be a whole number times $h\nu$ (although it wasn't really clear he believed wholeheartedly in this constraint). Nonetheless, in either view, the energy of any specific molecule was something that could and would vary in time, whatever the temperature of the surroundings. But each molecule is typically in contact with other molecules through collisions (in the gaseous state) or mutual vibrations in the solid state, so that if the material's temperature is fixed, each molecule has a definite *average* energy that *does* depend on the temperature of the surrounding environment and is termed the thermal energy.

Several times in his papers on statistical mechanics Einstein had already done a calculation for the kinetic part of the average thermal energy, the contribution of molecular velocity to the total thermal energy. It was a trivial generalization of those calculations to consider a molecule that vibrates back and forth and hence also has potential energy. Potential energy is just that—energy stored by doing work against a force. Continuing the spring analogy, when a mass on a spring is pulled and then clamped at a greater extension, potential energy is stored in the mass, which can be released when the mass is unpinned and oscillates back and forth. When a molecule vibrates, its chemical bonds are compressed and stretched like a spring, alternately storing and releasing potential energy. In this case, according to Maxwell/Boltzmann/Einstein statistical mechanics, the average potential and kinetic energies are equal and have a simple relation to temperature, $E_{mol} = kT$, where k is Boltzmann's constant (the same one that comes into the formula for the entropy, $S = k \log W$), and T is the temperature.[1] Note that this average thermal energy is *independent* of the frequency.

It may initially seem counterintuitive that two identical masses connected to springs that vibrate at very different frequencies (i.e., have different degrees of stiffness) can have on average the same energy. The high-frequency vibrating mass will oscillate faster, and thus will have higher kinetic energy, which, for an oscillator, also must be equal on average to its potential energy. Thus it should just have more total energy, right? No, because the two springs will also vibrate different

---

[1] Here and elsewhere I assume that temperature is measured using the absolute (kelvin) scale, where it is always a positive number.

distances from their unstretched positions. The relation $E_{mol} = kT$ tells us that the high-frequency molecule must make smaller-amplitude vibrations than the low-frequency molecule, in just such a way that the average energy of the two is equal. This statement of classical statistical mechanics, that all vibrating structures have the same average energy, has a fancy name: *the equipartition theorem*; it implies that the total energy of the system is equally shared by each microscopic part. This theorem will be very important in Einstein's reasoning going forward.

Since Einstein had by now developed a great understanding of classical statistical mechanics, he surely would have leaned initially toward the "obvious" equipartition formula: $E_{mol} = kT$, which via Planck's reasoning would immediately yield a hypothesis for the radiation law. However, as we shall soon see, using the hypothesis so obtained gives a paradoxical result, so Einstein at some point rejected this approach. Planck's route, on the other hand, involved an obvious fudge, the ad hoc hypothesis of a minimal energy element, with no fundamental justification at all. This dead end is the point at which ordinary mortals would have thrown up their hands and given up. Instead, in one of the greatest demonstrations of flexible thinking ever, Einstein abandoned his beloved classical statistical mechanics and opened his mind to a new and bizarre possibility, the possibility that the hallowed Maxwell equations, whose perfection he had long admired, were not the final word on the nature of light.

CHAPTER 9

TRIPPING THE LIGHT HEURISTIC

"The wave [Maxwell] theory of light . . . has proved itself splendidly in describing purely optical phenomena and probably will never be replaced by another theory. . . . [However] it is conceivable that . . . the [wave] theory of light may lead to contradictions with experience when it is applied to the phenomena of production and transformation of light. Indeed it seems to me that the observations regarding 'blackbody radiation,' photoluminescence, and [the photoelectric effect] . . . can be understood better if one assumes that the energy of light is discontinuously distributed in space. *According to the assumption to be contemplated here, when a light ray is spreading from a point, the energy is not distributed continuously over ever-increasing spaces, but consists of a finite number of energy quanta that are localized in points in space, move without dividing, and can be absorbed or generated only as a whole*" (italics added).

With these comments Einstein began his paper that presented a "heuristic point of view" (not a theory) on the production and transformation of light. A noted Einstein biographer has called the last statement of this quotation "the most revolutionary sentence written by a physicist of the twentieth century." There had been no significant work on a particulate view of light since Newton's theory of light as granular was definitively refuted in the early part of the nineteenth century. Thus Einstein's work has no real antecedents; it is a bolt from the blue. In contrast, what became special relativity theory was very much in the air by 1905, and when Einstein wrote his first paper on that topic, four months later, it involved a new derivation of mathematical properties of space and time that had already been written down by Lorentz (albeit without the radical interpretation given to them by Einstein).

The paper hypothesizing that light could be conceived of as a stream of particles (which Einstein called "quanta" and which we now refer to as "photons") is seventeen pages long and consists of an introduction and nine numbered sections. It is clearly written and would have been relatively easy to understand for an expert like Planck (who was theory editor for the journal). The paper makes a number of assertions, both qualitative and quantitative, about experimental phenomena that can be understood from the new point of view it puts forward but are puzzling when viewed from the vantage point of Maxwell's theory. Astoundingly, every statement in this paper is correct when assessed on the basis of our modern quantum theory of radiation and its interaction with matter.

While Einstein discusses experiments related to light extensively in the paper, this is not how he begins. Instead, he starts with the theoretical consideration already mentioned, that classical statistical mechanics fails to describe blackbody radiation. In this first section, titled "On a difficulty encountered in the theory of blackbody radiation," he explains, in a manner similar to Planck's, that to analyze the energy density of blackbody radiation, one does not need to assume statistical mechanics applies to thermal radiation. Instead, since the radiation is always exchanging energy with matter, one needs only to figure out the average energy per molecule of a gas in contact with electrons, which then emit light (here, ironically, he refers to the "sad specimen" of Drude and his electron theory for support). Having established this, he goes right to the point: classical statistical mechanics insists that the energy per molecule is kT. End of story. He then invokes Planck's own equation, which says that the blackbody energy density at frequency v, which he denotes by the Greek letter ρ, is proportional to this molecular energy, for which Planck had substituted, not kT, but a totally different answer, which he had gotten by working backward from his entropy calculation. But, Einstein points out, if one instead sticks with the answer $E_{mol} = kT$, the only answer that makes sense according to classical reasoning, there is a problem with the total energy of the radiation. Actually, a really big problem: the total energy summed over all the frequencies of radiation is infinite. Even though physicists toss around infinities nonchalantly, we are still not allowed to have an infinite answer for something you can measure. The total energy coming out of a finite-size blackbody had been measured; it obeys a known relation, called the Stefan-Boltzmann law,[1] and it is most assuredly finite. So the obvious approach gives an impossible result.

The source of the impossible answer was easy to trace. In a gas there is a very large but finite number of molecules; thus each molecule can have a fixed amount of energy, kT, and the total energy will be some large but finite quantity. In a box with trapped radiation, however, there are an *infinite* number of wavelengths of radiation that fit inside the box (remember that the wavelength of Maxwell radiation can be arbitrarily small). If each wavelength of radiation carries the same energy, then the total energy adds up to infinity, just as Einstein found. Is it possible that only some wavelengths actually get their share of the energy, leaving others below their thermal quota? No, it is not possible. Since the radiation and electrons are always interacting, if some wavelengths didn't get their "fair share" of the energy, then radiation would continually suck energy out of matter until all the matter cooled to absolute zero. Much later the physicist Paul Ehrenfest[2] came up with a catchy name for this phenomenon: *the ultraviolet catastrophe* (almost all the energy should flow into the shorter, ultraviolet wavelengths). Not to worry, it doesn't happen.

---

[1] This relation is that the total radiation energy flowing out of the surface of a black-body is $J = \sigma T^4$. It was derived by Boltzmann's teacher, Josef Stefan, and generalized by Boltzmann; σ is called the Stefan-Boltzmann constant and its value was known by 1905.

[2] Paul Ehrenfest was an Austrian-born Jewish physicist whom Einstein would meet in Prague in 1912. As Einstein recalled, "within a few hours we were friends, as if made for one another by our strivings and longings." Later that year Ehrenfest became a professor at Leiden and arranged regular visits there by Einstein for the next two decades.

This is the brick wall that Einstein had run into in trying to explain blackbody radiation from statistical mechanics. The only answer you can get from Newtonian mechanics and Maxwell/Boltzmann theory is not just wrong (i.e., out of agreement with experiment); it is absurd. This was why Planck had to twist himself into knots and introduce "energy elements" to get his answer. Einstein had major reservations about this gimmick, and had devised a means to make his conjectures about quanta of light without committing himself to whether there was any truth in Planck's notion.

Despite his reservations about Planck's method for deriving his radiation law, Einstein treats Planck quite delicately in his paper. One might have expected that the angry young maverick who had confronted Drude and Boltzmann about their supposed errors would have liked nothing more than to emphasize the shortcomings of Planck's approach in print. Planck had tiptoed to the very edge of the ultraviolet catastrophe when he noted that his smallest energy constant, h, could not be taken arbitrarily small (which would have restored the continuous nature of energy). If it were so taken, one would immediately encounter Einstein's absurd result for the total radiation energy; however, Planck had not seen fit to mention the menace he had narrowly escaped. Einstein had discovered that the kaiser Planck had no clothes, and who better than him to tell the world?

However, a cryptic comment by Besso much later, in 1928, sheds light on his change of tone: "On my side, I have been your public during the years 1904, 1905; by helping you to edit your communications on the problem of quanta, I deprived you of some of your glory; but on the other hand, I procured you a friend, Planck." It appears that Besso prevailed on Einstein to revise an earlier version of this paper in which more pointed comments were made about the correctness of Planck's derivation. Besso is on target: he probably did deprive Einstein of some of his glory; had Einstein been more direct in discussing Planck's work, his own role as the first person to propose seriously the quantization of energy, as a property of atomic mechanics, would have been much clearer.

At any rate, in the published version of the paper Einstein confines himself to the innocuous comment, "Planck's formula for $\rho(v)$ . . . has been sufficient to account for all observations made so far." Einstein is not trying to determine the correct radiation law here anyway, but instead his goal is interpreting the *meaning* of this law, which apparently *is* correct but, as he has just shown, conflicts with atomistic theory as then conceived. To do this he finds it sufficient to consider only the shorter wavelengths of thermal radiation, and to use for convenience the older Wien law, which fails for the longer wavelengths but actually fits the data very well in the short-wavelength, high-frequency range.[3] Assuming that Wien's law is *approximately* correct, he can work backward à la Planck and find the approximate entropy of blackbody radiation. He finds something very striking. The mathematical equation for this entropy is identical to that of a molecular gas, except that where the number, N, of gas molecules would appear in the formula for the molecular entropy, the expression E/hv appears for the blackbody entropy (where E is the total energy of the radiation of frequency v). Einstein makes the immediate connection: if light consists of a train of particles ("quanta") each of energy hv, then E/hv is the number of those quanta, and the blackbody entropy behaves exactly like a gas of independent molecules. In other words, the short-wavelength limit of the blackbody law suggests that light has particulate properties!

Of course this is not a proof. The only laws physicists believed at the time give the absurd answer for the total radiation energy (infinity) that he has already stated at the beginning of the paper. Thus Einstein doesn't know any more fundamental way of justifying his light-quanta hypothesis than through the mathematical analogy he has just given. However, this picture, even in the absence of the new and more accurate set of atomic laws that must underlie it, is enough to explain many puzzling experimental observations.

Here is how he was able to do this. First, he assumed (correctly) that whatever the new laws were, energy is still conserved. Energy conservation means that whenever some energy is given up by something (e.g.,

---

[3] In this context, high frequency means that energy element, hv, is larger than the classical thermal energy, $E_{mol} = kT$, and low frequency refers to the opposite situation.

a molecule or a light wave), that energy is transferred to some other physical object or process so that energy is not created or destroyed, just redistributed.[4] His new picture implied that light transfers energy to matter in a different way than in the Maxwell wave picture. In classical physics the energy in a wave is proportional to its intensity, which is proportional to the square of the maximum height of the wave. This makes a lot of sense intuitively. Waves are disturbances in a medium that transport energy from one region in the medium to another. For example, water waves bring energy into the shore from winds in the oceans. We all can see that a gnarly thirty-footer brings in more energy than a five-foot swell; we instinctively measure the power of waves by their height. This is exactly how the energy of a light wave is measured in the classical wave theory of electromagnetism, their "height" being the strength of their electric field.

Einstein was proposing a radical reimagining of this on the atomic level. He proposed that the light wave really is a train of particles (of indefinite but presumably very small size), the quanta, which, because of their localized nature, can interact and exchange energy only with individual atoms or molecules. The intensity or height of the light wave tells one *how many* quanta the wave contains in a certain region of space, but it doesn't determine the energy of each quantum of light. Instead, the energy of the quanta depends only on the frequency (or wavelength) of the wave according to the relation $\varepsilon_{quant} = h\nu$. This is just the Planck relation, but not for the energy of the molecular vibrations, rather for the quanta of light (photons).

Notice how strange this conclusion is from the point of view of the traditional wave picture. The frequency of a water wave just determines its wavelength (the spacing of the crests), which doesn't affect how much energy is transferred each time a crest crashes on the beach. Einstein agreed that higher-intensity light waves carry more

---

[4] Later that year Einstein would generalize this rule via the famous equation $E = mc^2$. Sometimes energy can be transformed into mass and is "lost," although in principle it can always be transformed back, so that mass-energy is still conserved. Such effects are typically negligible with absorption and emission of light, although essential in nuclear reactions.

total energy (because they consist of more quanta), but carrying more quanta matters little for how they interact with a *single* molecule or atom. That is because it is very improbable that two quanta will "meet" at the same point at the same time and thus be able to transfer twice as much energy to the same molecule.[5] So, in his new picture, what can happen in such a transfer is controlled by the frequency of the light and not its intensity.

Suddenly very puzzling observations made perfect sense. Einstein's first example was a phenomenon involving the absorption of light known as the Stokes rule. George Gabriel Stokes, born in the Irish county of Sligo in 1819, was the first of the three great Cambridge mathematical physicists[6] of the nineteenth century, along with Maxwell and William Thomson (later named Lord Kelvin). One of his most famous discoveries was that certain substances, when they absorbed light would "fluoresce"; that is, they would reemit the light, not as blackbody radiation, for which the wavelength is determined by temperature, but rather as visible light. However, Stokes noted that the reemitted light was always of a lower frequency (longer wavelength) than the absorbed light: this is the Stokes rule. Most dramatically, many substances could absorb ultraviolet (invisible) incident radiation and shift it down to lower frequencies in the visible range (this effect is now used as a key diagnostic probe in modern chemistry and biochemistry). Ironically, Stokes never published his famous rule;[7]

---

[5] While this is improbable for normal beams of light, with the particularly intense beams generated by modern lasers it becomes much more likely, leading to so-called nonlinear optical effects. An important new form of microscopy, known as two-photon microscopy, which gives superior imaging of living tissue, is based on this new possibility. Einstein is aware of and mentions the possibility of nonlinear processes involving several photons in his article but assumes, correctly, that they are rare.

[6] None of the three were English: Thomson was born in Belfast in 1824 and Maxwell in Edinburgh in 1831. Both Stokes and Thomson lived into the twentieth century, whereas Maxwell, to the great detriment of science, passed away in 1879. Maxwell, now regarded as the greatest of the three, was the only one not to be knighted.

[7] Lord Rayleigh, the successor to these three leaders, commented that Stokes may have suffered from a "morbid dread of mistakes," which inhibited publication.

instead Thomson announced it in 1883 but gave Stokes full credit for the discovery.

This rule makes no sense from the classical wave point of view. Since the energy in a classical light wave has nothing to do with its frequency, why can't a conversion of light to higher frequency sometimes happen? After all, the frequencies at which molecules emit light presumably have something to do with how tightly the atoms are bonded together, how the molecules vibrate, and so on. If the light wave is dumping energy in and it is being reemitted as a new light wave, then shouldn't it come out at whatever the natural frequencies of emission are for that molecule, independent of whatever frequency is used to dump the energy in? (Perhaps this obvious conundrum deterred Stokes from publishing, since there was no reasonable explanation for his rule.) Einstein's quantum picture handles this with ease.

If the best a molecule can do is absorb all the energy in *one* quantum of light, then the most energetic quantum it can reemit will have at most the same energy, and hence the same frequency; however, absorption almost always is accompanied by some amount of energy lost to the molecule (a sort of molecular friction), so in fact the highest energy and hence the highest frequency that can be reemitted is always lower. Einstein realizes that this conclusion is quite general: "it makes no difference by what kind of intermediate process this end result [the emission of a photon] is mediated. If the photoluminescent substance is not to be regarded as a permanent source of energy, then, according to [conservation of energy], the energy of the produced energy quantum cannot be greater than that of the producing energy quantum, hence we must have [the produced frequency less than or equal to the producing one]. This is the well-known Stokes rule."

This was a very nice first flexing of Einstein's quantum muscles, but it was still just a qualitative explanation for something that was already known. His idea was so radical that it would take something more to get people to take it even a little bit seriously. Fortunately, he comes up with a more dramatic example and quantitative prediction in the next section of the paper. The great German physicist Heinrich Hertz, the first demonstrator of Maxwell radiation, had also discovered by accident a phenomenon that would ultimately spell the downfall of pure Maxwellian electrodynamics. This phenomenon is called the "photoelectric effect."

When light, particularly blue or ultraviolet light, is incident on and absorbed by a substance, yet another process can occur (besides the emission of blackbody radiation or the fluorescence of lower-frequency Stokes radiation). Sometimes the substance can eject an energetic "cathode ray," which by that time had been identified as a speeding electron. These fast electrons can be collected and run through a circuit to generate a "photocurrent." The same principle is operative in all modern solar cells, except that the electrons are not ejected from the material but are just promoted to a higher energy state, where they move much more freely and can be extracted as electrical energy in a circuit. By exactly the same kind of reasoning he had applied to the Stokes rule, Einstein explained the photoelectric effect. He analyzed the consequences of conservation of energy, combined with the principle that only *one* quantum of light can interact with *one* electron at a time.

First, it always takes some energy to knock electrons out of a solid. The electrons are trapped by their attraction to positively charged atoms and also by the surfaces of the material, so they need to absorb extra energy to be kicked out into space. This energy is pretty large on the atomic scale, comparable to the amount of energy in one quantum of blue or even ultraviolet light. So if one bathes the material in a beam of red light (no matter how intense), since each quantum of red light is individually too feeble, *no* photoelectrons are produced. Again, this is totally baffling from the classical point of view. From that perspective, the more intense the red beam is, the more energy it has, and the more likely it should be to produce photoelectrons. But increasing the red beam's intensity does nothing. In contrast, a rather weak beam of ultraviolet light does produce photoelectrons. A paradox for Maxwell, but not in the quantum picture—one solitary ultraviolet photon, with its higher frequency, has enough energy to do the job. So this observation is immediately explained by Einstein's light quanta.

But he easily goes further. The ejected photoelectrons are winging their way toward the collector, which is a metal contact absorbing electrons and delivering them to the attached circuit. If this collector is charged with a large enough negative voltage, the electrons are repelled and the photocurrent ceases. A simple further application of conservation of energy allows Einstein to predict that this "stopping voltage" is precisely proportional to the frequency of the light. In other words, if you make a graph of stopping voltage versus frequency, it is a perfectly straight line with a specific value for its slope. And what is that value? It is exactly Planck's constant, h, for any material! Now that was a strong and precise prediction.

Moreover, it was an interesting and surprising prediction because it moved Planck's constant, h, from the arena of thermal radiation to the seemingly unrelated area of photoelectric phenomena. It strongly suggested that this constant was of general significance for physics. This is just the kind of thing experimental physicists love: an important new idea connected to a clear and easily falsified prediction. The statement that stopping voltage must have a linear dependence on the light beam's frequency, independent of its intensity, was bizarre enough. But also the graph of this line must have a universal, material-independent slope? It was too good to be true. Surely someone would jump on this.

Except . . . the photoelectric experiments required were very difficult, and Einstein, a virtual unknown who was contradicting the wave theory of light, had hardly more credibility than a crackpot, whose writings were to be thrown in the wastebasket. And unfortunately the existing data did not help very much. They did appear to show the strange dependence on frequency and independence of intensity that Einstein's picture explained, but they were nowhere near accurate enough to verify or contradict his universal-slope prediction. He was only able to make a rather hopeful statement: "As far as I can see, our conception does not conflict with the properties of the photoelectric effect observed by [the German physicist Philip Lenard]." Einstein would have to wait several more years for his quantum of fame.

CHAPTER 10

ENTERTAINING THE CONTRADICTION

"I do not seek the meaning of the quantum of action (light quantum) in the vacuum but at the sites of absorption and emission, and assume that processes in vacuum are described exactly by Maxwell's equations." This was Max Planck's first known response to Einstein's heuristic theory of light quanta, sent to Einstein in a letter of July 6, 1907, more than two years after the publication of the "revolutionary" paper of 1905. Planck must surely have known of Einstein's ideas much earlier, since he was the theory editor of the journal in which they were published, *Annalen der Physik*. Unfortunately none of the referee or editorial comments on Einstein's papers of 1905 has survived, so we don't know how direct a role Planck had in approving them for publication. Planck was known for being open to publishing scientific contributions with which he disagreed, as long as they did not contain clear errors, and this tolerance likely came into play with Einstein's paper. Planck consistently opposed the idea of the particulate nature of light in vacuum, in a respectful but firm manner, for at least a decade after its publication. And he was not alone in this attitude; among the eminent physicists of the time, only Johannes Stark, whose outstanding work on the photoelectric effect Einstein had mentioned, openly supported Einstein's radical new view of light. In May of 1909, in response to a letter from Einstein, Lorentz himself sent a lengthy, technical reply to Einstein, detailing what he saw as the insuperable problems with his hypothesis. He concludes, "It is a real pity that the light quantum hypothesis encounters such serious difficulties because otherwise the hypothesis is very pretty, and many of the applications that you and Stark have made of it are very enticing."

What were the "serious difficulties" of which Lorentz spoke? They arose because of the conflict between two basic categories in the physics of the time: waves and particles. Newton's laws had introduced the idea of mass, the quality of a material that resists change of motion and that responds to and generates gravitational attraction. Although atomic theory was in its infancy, the idea that everyday massive objects were made of smaller, more fundamental building blocks (i.e., atoms and molecules) had been widely used by physicists since the time of Maxwell and was winning the day by 1905. Atoms were the fundamental particles of physics, although soon to be understood as made up of protons, neutrons, and electrons. (Modern physics has added many other particles to this classification and further subdivided the atomic nucleus into quarks). So the idea that macroscopic "stuff" (solids, liquids, and gases) is an aggregate of atomic-scale particles was commonplace at the time of Einstein's early work.

Moreover, it was clear that when these particles aggregate in large quantities, such as in an ocean or in the atmosphere, they create media in which disturbances can propagate, disturbances that in these cases are called water waves or sound waves. It is important to realize that the particles of water or air are the *substrate* in which the waves propagate; it not possible to generate such "classical" waves without their substrate (in space no one can hear you scream). In the classical view, the fundamental objects are the particles that make up the medium, and the waves are derivative objects. Moreover, it is critical to understand that these waves are collective phenomena; waves move through the medium, but the particles don't. Waves are not a whole bunch of particles moving in the same direction.

This is illustrated nicely by a new kind of wave, discovered sometime in the 1980s, which we will refer to as "fan waves." In the classical fan wave the particles are the sports fans in a stadium, who, because of boredom or some other stimulus, spontaneously generate collective motion. In an ideal, fully developed, clockwise fan wave all the fans from the first to last row in the upper deck stand up and then sit down in a brief period (about two seconds), causing, by poorly understood interactions, the fans immediately next to them on their left to do the same thing immediately thereafter. This disturbance in the crowd then propagates around the stadium, creating a nice visual effect, until it is damped out either by loss of synchronization or loss of interest. Anyone who has contributed to such a wave realizes that the particles (the fans) do not move in the direction of the wave; they just bob up and down. It is the disturbance, the wave, that propagates, not the "particles" of the medium.

To this extent fan waves are typical classical waves in a medium. To complete the analogy, however, we will have to embellish a bit on the conventional fan wave to allow for a further critical feature, interference of waves. Imagine that all the fans are standing already (it is a particularly exciting moment in the game) and can make two different kinds of waves, by either raising their arms above their heads or lowering them down to their knees, as in a revival meeting. Also allow for the possibility that waves can go either clockwise or counterclockwise. You look to your left or right, and if the fan next to you raises his hands, you do the same; if he lowers them you do the same as well. Now some wise guy starts a clockwise wave by raising his hands, and his friend behind him starts simultaneously a counterclockwise wave by lowering his hands. These two waves propagate around the stadium in opposite directions at the same speed, and so halfway around they meet. At the column where they meet, the fans on the right raise their hands just as the fans on the left lower theirs: the fans in the middle don't know what to do. So they do nothing. The two waves have met up, "out of phase" as the physicists say, and they cancel each other out.

This is a somewhat fanciful illustration of the interference of waves, which are extended disturbances in a medium, having both an amplitude (how big the wave is at any given point) and a phase (how close the wave is to a peak or a trough at any given point). Depending on their phases, waves interfere and are larger where the crests coincide and smaller (or zero) where a crest and peak coincide. This is the sine qua non of a wave. But note that when we have destructive interference and two waves cancel out, the particles of the medium are still there (i.e., the fans in our example); they are just undisturbed. Waves are disturbances, so they can be positive or negative and can cancel each other out; you can add one and one and get zero. Particles cannot. (Two fans claiming a single seat *will* create destructive interference, but of a different kind.)

This was how all waves were conceived of until 1905. However, a major challenge to this understanding was implicit in Maxwell's discovery of electromagnetic waves. Here the propagating disturbance was an electric and magnetic field, but there was no obvious medium in which it could propagate. Physicists since the time of Newton had hypothesized that heat and possibly even light propagated through a transparent medium known as the "ether." Maxwell's discovery now confirmed its existence and that it was the substrate through which all EM waves propagated.

The absolute necessity for such a medium was so evident that Heinrich Hertz, the first to demonstrate reception and transmission of radio waves, expressed it thus: "Take electricity out of our world and light vanishes; take the luminiferous ether out of our world and electric and magnetic fields can no longer travel through space."

This medium was, however, highly problematic. Despite its stubborn invisibility, it had to be all-pervasive, since apparently EM waves could propagate everywhere. It couldn't have much (if any) mass, because it would then have gravitational effects, which were not in evidence. And since the earth moves in different directions at different times of year, the velocity of light on earth should vary in some manner, just as a water wave appears to move more slowly to a boat moving in the same direction. However, experiments testing the speed of light showed no hint of this effect.

But what choice did one have other than postulating an ether? Try creating a fan wave in an empty stadium. You don't have to be an Einstein to see that you can't have the wave without the medium. It turns out, however, that you *do* have to be Einstein to suggest that you *can* have the wave without the medium.

As noted above, the familiar waves of classical physics, which propagate as a disturbance in a medium, look different to an observer moving with respect to that medium. The surfer on the crest of a water wave sees an almost stationary wall of water roiling around him. By the same logic the young Einstein, in his school days at Aarau, had imagined moving along next to a light wave at the speed c and seeing a stationary electric field that no longer oscillated. This apparently conceivable physical situation made no sense to him: "But such a thing does not seem to exist, either on the grounds of experience or according to the Maxwellian equations." The leading theorists of the time, Lorentz and the French mathematical physicist Henri Poincaré, grappled with this conundrum and, while making major mathematical advances, kept the physical picture of electromagnetic waves tied to the ether. Einstein also pondered this puzzle on and off during his student years and afterward, and it was finally in May of 1905, two months after he had submitted his paper on light quanta, that the answer came to him: if time itself were not absolute but "flowed" differently for observers in uniform relative motion, then all the apparent contradictions could be resolved!

This was the key idea of Einstein's "rough draft" on the "electrodynamics of moving bodies which employs a modification of the theory of space and time" that he spoke of in his vivacious letter to Habicht in May 1905. Within two months this idea has been developed into his famous paper on what became known as the special theory of relativity. This work has received much attention in the literature, and we will not review it here except to quote part of one critical sentence: "The introduction of a 'light ether' will prove superfluous, inasmuch as . . . no 'space at absolute rest' endowed with special properties will be introduced."

It is important to understand that the theory of special relativity is completely independent of quantum theory and can be seen (along with the later general theory of relativity) to be the culmination of classical physics and its deterministic worldview. Special relativity makes sense of classical Maxwellian electromagnetic waves, and it does so without the somewhat embarrassing, unobservable ether. Two months after undermining Maxwell's equations with his heuristic theory of light quanta, he vindicates them from a host of experimental challenges by banishing the ether. Talk about creative tension.

On the other hand, by getting rid of the ether, it was clear that Einstein was now prepared to accept waves that *do not travel in a medium*, "fundamental waves." A few years later he made this thinking explicit: "one can obtain a satisfactory theory only if one drops the ether hypothesis. In that case the electromagnetic fields which constitute the light will no longer appear to be states of a hypothetical medium, but rather independent entities emitted by the source of light." Since EM waves were, from this point of view, a completely new kind of entity in physics, perhaps they *could be* something in between a classical particle and a classical wave, as suggested by Einstein's notion of light quanta. En masse they exhibited interference like classical waves and hence could cancel one another out, but when exchanging energy with matter they acted like localized particles. Einstein was willing to entertain this contradiction. He, of all the physicists of his time, was the only one to really imagine that these two apparently conflicting concepts could be married. Over the next six years Einstein would devote the main part of his energies to consummating this difficult marriage.

CHAPTER 11

STALKING THE PLANCK

"The three of us are fine, as always. The little sprout has grown into quite an imposing impertinent fellow. As for my science, I am not all that successful at present. Soon I will reach the age of stagnation and sterility when one laments the revolutionary spirit of the young. My papers are much appreciated and are giving rise to further investigations. Professor Planck (Berlin) has recently written to me about that."

Thus the twenty-seven-year-old Einstein wrote to his former Olympia Academy colleague Maurice Solovine in April 1906, in the aftermath of his miracle year. The "little sprout" he spoke of was his first son,[1] Hans Albert, who had been born on May 14, 1904, and was now coming up to his second birthday. At this point Einstein remained virtually unknown to the wider physics community, having personally encountered only a handful of physicists during his studies at the Poly and in subsequent research, none of whom were eminent theorists. He was working steadily eight hours a day, six days a week at the patent office, describing himself (with characteristic saltiness) to a friend as "a respectable Federal ink pisser with a decent salary." His salary had recently grown a bit more respectable. In April of 1905 (in the midst of his creative epiphanies) he had again submitted a dissertation for a PhD to the University of Zurich, choosing his safest work of the mo-

---

[1] Hans Albert was Einstein's second child; his first, a daugher named Lieserl, was born to Mileva in Novi Sad, Serbia, in 1902, before their marriage. The child was left in Serbia with friends or relatives under unclear circumstances, and her subsequent history is unknown, although it is believed that she did not survive to adulthood.

FIGURE 11.1. Albert Einstein in 1904 with his wife, Mileva Maric, and his young son, Hans Albert. ETH-Bibliothek Zurich, Image Archive.

ment as the topic, that on irregular molecular movement ("Brownian motion") and the determination of Avogadro's number. This time Kleiner and the committee accepted the thesis, and as a consequence he was promoted to technical expert second class at the patent office, with a 15 percent increase in salary.

While Einstein personally was unknown to the great men of theoretical physics, his work had already, barely a year later, made a significant impact. As the above quotation makes clear, Planck had already written to tell him his work was very much appreciated (although Planck's letter has not survived). One might have supposed that Einstein's work on light quanta, relating as it did to the central achievement of Planck's career, the blackbody radiation law, would have been the main object of Planck's attention and appreciation. However this was not the case. Nothing is known of Planck's reaction to Einstein's quantum hypothesis until his letter of July 1907, quoted above, definitively rejecting the idea of light quanta in vacuum. In contrast, Planck immediately embraced the special theory of relativity; he, not Einstein, gave the first public lecture on the subject (crediting Einstein of course) in the fall of 1905 shortly after the theory was published in September of that year. (As the theory editor of *Annalen der Physik*, Planck naturally would have seen the paper when it arrived at the end of June.)

Not only did Planck quickly publicize relativity theory, but he also paid it the highest compliment possible from a working scientist: he redirected his research to the study and extension of the theory. In 1906 he published the first major contribution to relativity theory not due to Einstein, a proof that relativistic mechanics was compatible with the "principle of least action,"[2] an alternative mathematical formulation of classical mechanics that was flexible enough to encompass the alterations of Newton's laws required by relativity theory. From 1906 to 1908 all of Planck's new research related to relativity theory. Because of Planck's stature in the field, his immediate "lively attention" to relativity theory endowed it with a credibility and importance that it otherwise might not have achieved for some time. Einstein acknowledged this in a tribute to Planck in 1913 when he stated, "It is largely due to the determined and cordial manner in which [Planck] supported this theory that it attracted notice so quickly among my colleagues in the field."

---

[2] The mathematical quantity called "action" mentioned here is the same one that led Planck to call his constant, h, the "quantum of action" because it has the same physical "units" of mass times velocity times distance. In classical physics action is not quantized, and Planck did not invoke the constant h in his work on relativity theory.

That Planck reacted in this manner was completely consistent with his personality and philosophy of science; he recognized that relativity theory, as strange as it appeared to laypeople and to some physicists, in fact *completed* classical mechanics and made it compatible with Maxwell's electromagnetic theory. Einstein himself described it as "simply a systematic development of the electrodynamics of Maxwell and Lorentz." He frequently emphasized the continuity of relativity theory with earlier physical principles: "There is a false opinion widely spread among the general public that the theory of relativity is to be taken as differing radically from the previous developments in physics. . . . The four men who laid the foundations of physics on which I was able to construct my theory are Galileo, Newton, Maxwell, and Lorentz." Special relativity was a theory a purist like Planck could love. Wayward quanta of light, propagating in vacuum but still interfering like waves, impugning the integrity of Maxwell's equations; now that was a completely different matter. The best he could do was to forgive this impetuous genius a youthful indiscretion.

Planck finally did address the light-quantum hypothesis in the letter of July 1907 quoted above ("[I] assume that processes in vacuum are described exactly by Maxwell's equations"), but apparently only in response to repeated prodding by Einstein, whose own preceding letters to Planck have been lost. After stating his belief in the validity of Maxwell's equations, and that the "quantum of action" (h) pertained only to the exchange of electromagnetic energy with matter, he continued, "but more urgent than this *surely rather old question* is at the moment the question of the admissibility of your relativity principle . . . as long as the proponents of the principle of relativity constitute such a modest little band as is now the case, it is doubly important that they agree among themselves" (italics added). Seven years after slipping discontinuity into physics through the back door, Planck still did not see this for the epoch-making event it was. In contrast, by the end of December 1906, Einstein had already realized that Planck's quantum of action was not going to remain trapped in Pandora's radiation cavity, and that the challenge it presented to the worldview of physicists was more fundamental than that of relativity theory.

As mentioned above, in his 1905 paper on light quanta Einstein had sidestepped a direct confrontation with Planck by crediting the Planck radiation law with being "sufficient to account for all observations made so far" but basing all his conclusions on the Wien approximation to it, which is valid for short-wavelength radiation. His resolute insistence that the statistical mechanics of the time could only give an impossible answer—kT of energy for each allowed wavelength in the cavity, leading to the ultraviolet catastrophe—indicates that at that time Einstein regarded Planck's "derivation" of his radiation law, employing the artifice of the energy element, hv, as highly suspect if not downright incorrect. By March of 1906, almost exactly a year after publishing his revolutionary light-quantum hypothesis, he had apparently reconsidered this view, and submitted a paper arguing that the Planck formula *requires* the concept of light quanta.

> In a study published last year I showed that the Maxwell Theory of electricity in conjunction with the theory of electrons leads to results that contradict the evidence on black-body radiation. By a route described in that study I was led to the view that light of frequency v can only be absorbed or emitted in quanta of energy[3] hv. . . . This relationship was developed for a range that corresponds to the range of validity of Wien's radiation formula.
>
> At the time it seemed to me that in a certain respect Planck's theory of radiation constituted a counterpart [alternative] to my work. New considerations, which are being reported [here], showed me, however, that the theoretical foundation on which Mr. Planck's radiation theory is based differs from the one that would emerge from Maxwell's theory and the theory of electrons, precisely because Planck's theory makes implicit use of the aforementioned hypothesis of light quanta.

These are the opening words of the 1906 paper. Note here the consistency with his 1905 paper, beginning first with the statement that

---

[3] Einstein did not actually use the symbol h introduced by Planck for his constant, preferring instead to use a ratio of older constants equal to Planck's constant. Einstein continued this practice for several more years, perhaps indicating a reluctance to accept h as a fundamental constant of nature.

conventional theory leads to a blackbody radiation law in contradiction to Planck's law (and for good measure he restates this incorrect law, which led to the ultraviolet catastrophe in the second section of the 1905 paper). Then he reiterates that his quantum hypothesis was based only on the Wien limit and not on the full Planck law. Finally he explicitly states that when he wrote his 1905 paper he believed that there was a tension, if not an outright contradiction, between the Planck law and the heuristic theory of light quanta. What had changed his mind on this?

Planck had introduced the quantization of energy for the "molecules" (he called them resonators) in his blackbody as a counting device, leaving it quite unclear as to whether this was a physical hypothesis or a mathematical convenience. From this hypothesis he derived the entropy of the resonators and then by further manipulations determined the distribution of energy among frequencies of radiation in the body. Now, in the 1906 paper, Einstein starts from exactly the same equation as Planck, relating the energy of radiation at frequency v to the average energy of each molecular resonator in the black body. In 1905 he had calculated the average oscillator energy by conventional statistical mechanics and got the answer that each one had the same energy, kT, which when transferred to radiation and added up over an infinite number of possible wavelengths gave infinity. Now he decides to tinker with conventional statistical mechanics. He writes a mathematical expression for the entropy of one of Planck's resonators, one that he (and independently Gibbs) had found several years earlier, which looks different from Boltzmann's famous $S = k \log W$ but which he shows is mathematically equivalent. In his new expression, instead of counting states, one finds the entropy by adding up contributions from all the possible energies of each resonator.[4] He finds that when he allows the energies to take continuous values, as they do in Newtonian physics, he gets an expression for the entropy that leads to the ultraviolet catastrophe. But if he simply uses Planck's restriction, that

---

[4] The method, very common in modern treatments, is to relate the entropy to the "free energy," which itself is obtained from the "partition function," a sum over functions of the energy of the system.

energy can increase only in steps of hv, the Planck law follows in a few steps of algebra. He then puts his cards on the table:

> Hence we must view the following proposition as the basis underlying Planck's theory of radiation: The energy of an elementary resonator can only assume values that are integral multiples of hv; by emission and absorption, the energy of a resonator changes by jumps of integer multiples of hv.

This in itself is not much of a mathematical advance over Planck; Einstein has just rearranged the mathematical route to Planck's formula in a way he finds more congenial and intuitive. To Einstein this approach makes it clear that the quantization of energy is not just a mathematical convenience but a hypothesis about nature, and one very closely allied to his hypothesis of quanta of light. Anyone who had read Planck's derivation carefully might have realized the same thing. In fact, however, we know of no other physicist of the time who *did* realize this, except for the omniscient Lorentz, and as we shall see, Lorentz ultimately drew the wrong conclusion from this realization—that the Planck formula must be wrong. Einstein does not even entertain the possibility that the Planck formula might be wrong—from his earliest work on the subject he seems to have accepted this law as an established experimental fact that must be dealt with (a view held by a knowledgeable few in Germany, but not more generally). Instead he attempts, as he did earlier in 1905, to find the law's *meaning* by linking it to the quantum hypothesis, but he is unsparingly honest about the flaws in his and Planck's reasoning. The starting point for Planck, and for him in the current paper, is a mathematical relationship between the energy of thermal radiation and the average energy of matter (resonators) in contact with that radiation, *a relationship that was found on the basis of assuming the validity of Maxwell's equations.* To then deduce the Planck law, one inserts a quantum hypothesis that is foreign to, and apparently contradicts, Maxwell's theory—hardly a compelling chain of logic.

Einstein continues: "For if the energy of a resonator can only change in jumps, then the mean energy of a resonator in a radiation space cannot be obtained from the usual theory of electricity, because the latter does not recognize *distinguished* [quantized] energy values of a resonator. Thus the following assumption underlies Planck's theory: although Maxwell's theory is not applicable to elementary resonators, nevertheless the mean energy of an elementary resonator . . . is equal to the energy calculated by Maxwell's theory." In other words, Planck uses Maxwell when it suits his purposes and doesn't use him when it doesn't. But Einstein is not pointing a finger at Planck; instead he is pointing out the need for a fundamental revision of physics to encompass quanta. "In my opinion the above considerations do not at all disprove Planck's theory of radiation; rather, they seem to me to show that with his theory of radiation Mr. Planck introduced into physics a new hypothetical element: the hypothesis of light quanta."

Herr Planck, however, declined to take credit for this outlandish new hypothesis, by all indications immediately upon learning of it, and for certain in his letter of July 1907, when he urged Einstein to focus on the very important issues of relativity theory and to put the potentially lethal case of the contradictory quanta in quarantine. Einstein did not take his advice; on the contrary, he was soon to spread the infection from radiation to matter. But while Einstein was plotting his next radical step, the great men of the field were just waking up to the dangerous paradox of thermal radiation within classical physics, the ultraviolet catastrophe. The eminent British physicist Lord Rayleigh had peeked at it six years earlier but had then averted his eyes; now it was becoming impossible to ignore.

CHAPTER 12

CALAMITY JEANS

In June of 1900, six months prior to Planck's historic act of desperation and long before Einstein's 1906 clarification of its meaning, John William Strutt, Lord and Third Baron Rayleigh, had noticed that something must be wrong with the Wien law. This was the law that was believed to describe blackbody radiation but would soon be found experimentally to fail at long wavelengths.

Lord Rayleigh, a member of the British nobility, had overcome his family's disapproval of the plebeian study of nature to ascend to the very pinnacle of British science. He had been a sickly youth, bouncing from school to school, and thus did not exhibit his talents early, even being turned down for a minor scholarship at Cambridge. Unexpectedly, he had blossomed at Cambridge, winning the top mathematics award ("First Wrangler") in 1865 under the exacting supervision of the formidable mathematics teacher E. J. Routh.[1] Upon graduation he ignored his father's reservations about a peer joining academia and won a fellowship at Trinity College, Cambridge. In 1871 his theory of the scattering of light waves by pointlike particles provided the first rigorous explanation of the blue color of the sky, and in 1877 he published his masterpiece: *The Theory of Sound*. By 1879 his reputation as a physicist was so great that he was chosen to succeed Maxwell as the head of the Cavendish Laboratory at Cambridge. After great success in this role he resigned in 1884 to concentrate again on his own research,

---

[1] Routh was known to be so sparing in his praise that when another First Wrangler, Lord Fletcher Moulton, produced an almost unheard-of perfect exercise, the only comment on the paper from Routh was "Fold neatly."

which spanned not only acoustics and electromagnetism but also arcana such as the "soaring of birds" and the "irregular flight of tennis balls." At that time he expressed his vocation thus:

> the domain of natural sciences is surely broad enough to satisfy the wildest ambition of its devotees. . . . Increasing knowledge brings with it increasing power, and great as are the triumphs of the present century, we may well believe that they are but a foretaste of what discovery and invention have yet in store for mankind. . . . The work may be hard, and the discipline severe; but the interest never fails, and great is the privilege of achievement.

By 1900, when he enters our story, he had in the previous year received the Copley Medal of the Royal Society, the same high honor it would award to J. W. Gibbs in the following year, and was four years from becoming the fourth recipient of the Nobel Prize in Physics.[2] The great generation of Maxwell, Stokes, and Thomson (Lord Kelvin) having relinquished the stage, Rayleigh was now the voice of British physics.

Rayleigh had been following the investigations of blackbody radiation in Germany and was aware that the Wien law was currently favored both on the basis of experiments and from extensions of the theory (Planck's extensions, in fact, before his famous retraction in October 1900). But he was unimpressed with the theory leading to the Wien law, dismissing it as "little more than a conjecture." In fact he had noticed that the mathematical formula expressing this law had an odd feature. Both the incorrect Wien law and the correct Planck law have the feature that at a given temperature there is a particular frequency of radiation that carries the largest fraction of the radiated energy; that is, both mathematical functions have a "peak" at a frequency that depends on the temperature of the body.[3] This peak frequency increases with increasing temperature, residing in the infrared at normal temperature but moving into the visible when matter is heated to six thousand

---

[2] He was recognized for his discovery of the element argon in the atmosphere.

[3] See appendix 2 for a graph showing the three radiation laws as a function of frequency.

degrees kelvin, as at the surface of the sun. Rayleigh was interested in how the peak frequency shifts upward when the temperature is increased. This could happen in two ways, either by the radiation energy increasing only at higher frequencies than the previous peak, or by the energy increasing *more* at the higher than at the lower frequencies, so that while the entire curve of energy output moves upward, its peak also shifts to higher frequency. Rayleigh's intuition was that a rising tide lifts all boats; that is, increasing the temperature of the body should increase the radiation energy output at all frequencies, so he thought the second option had to be the correct one. But he noticed that this was not the way the Wien law worked. The Wien law predicts that if one heats an object and measures the low-frequency side of the curve (below the peak), the energy at these frequencies will *stop* increasing once the temperature is high enough. Rayleigh was sure that this had to be wrong (and he was right; the correct, Planck law does not have this property).

Rayleigh was very familiar with the fact that the thermodynamic energy of a particle in a gas is simply proportional to its temperature. This fact is closely related to the equation $E_{mol} = kT$ (where k is Boltzmann's constant) for a simple vibrating molecule, emphasized by Einstein in both his 1905 and 1906 papers on quanta of light. As already mentioned, the classical prediction that all vibrating structures at the same temperature should have the same thermal energy, kT, is known as the equipartition theorem. The term "theorem" is a bit of a misnomer, since the soon-to-emerge quantum theory would disprove this statement, as least in the case of molecular vibrations. But at the time Rayleigh was first studying blackbody radiation, there was not yet even a hint of such drastic measures, so he was willing to rely on the standard classical statistical mechanics. He realized that the equipartition result, $E_{mol} = kT$, for vibrating structures of different frequency was just one example of the equal sharing of thermal energy. For a gas of freely moving particles, again each particle should have the same amount of thermal energy, but in this case the amount is different than for an oscillator at the same temperature:[4] $E_{gas} = 3kT/2$.

---

[4] For a gas particle moving in a single dimension, $E_{gas} = kT/2$, half that of the oscillator. This is because the gas particle has only kinetic energy and no potential energy,

As suggested by its name, classical statistical mechanics leads to the conclusion that all the thermal energy in the environment is equally shared among similar types of atomic motion. Rayleigh referred to this as "Maxwell's doctrine of partition of energy," and he had validated it in detail, using Newtonian mechanics, in a long paper written only a few months earlier in 1900. Such perfect energy sharing *would be correct* if molecules were made up of little Newtonian billiard balls, but it turns out that this isn't the world we live in, and equipartition doesn't hold in our quantum world.

This is a good thing. Like many egalitarian principles, equipartition can have unanticipated consequences. In this case the consequence would have been the ultraviolet catastrophe pointed out and rejected by Einstein in his paper on light quanta five years later. Unknown to Einstein, Rayleigh had been there first, but he didn't raise the alarm.

As Rayleigh pondered the Wien law in June of 1900, he drew on his great expertise in acoustics, which told him that the classical equipartition principle didn't really describe well the actual behavior of gases. The speed of sound in a gas (or mixture of gases like the atmosphere) depends on how its energy content changes with temperature, which is termed the "specific heat" of the gas. The properties of the specific heat of many gases did not agree with the equipartition concept; many gases had smaller specific heats than they should, if one just counted how many types of vibrations they were expected to have and assigned to each its equal share of energy. Rayleigh, noting this, opined, "what would appear to be wanted is some escape from the destructive simplicity [of the equipartition theorem]." The Great Escape, quantization of energy, would be found by Herr Planck in less than a year, but Rayleigh had no way of knowing this. He did know that equipartition was not working in general, and so he was reluctant to trust it in considering the blackbody law, which he charmingly termed, in his new paper of June 1900, the "law of complete radiation." Nonetheless he

---

which for the oscillator gives an equal contribution, $E_{mol} = kT/2 + kT/ = kT$. However, for the realistic case of a gas particle moving in all three directions in space, $E_{gas} = kT/2 + kT/2 + kT/2 = 3kT/2$.

realized that this equipartition principle could fix the odd feature of the Wien law he had noticed.

While admitting that "the question is one to be settled by experiment . . . in the meantime," Rayleigh wrote, "I venture to suggest a modification [of the Wien law] which appears to me more probable *a priori*." First he assumed that the radiation in the blackbody cavity could be thought of as the vibrations of an elastic medium, much like sound waves, essentially assuming a mechanical ether to support radiation (as did all the other physicists of his time). He termed each of these elastic vibrations, with different frequencies, "radiation modes." He then used equipartition to find the energy of those vibrations, but *only* for the low-frequency modes: "although for some reason not yet explained [the 'Maxwell-Boltzmann doctrine of the partition of energy'] fails in general, it seems possible that it may apply to the graver [lower-frequency] modes."

With this assumption he derives essentially the same expression for the classical blackbody law that Einstein finds in section 1 of his 1905 paper, except . . . he fudges it. Seeing that this would lead to an absurd result, infinite energy when summed over all wavelengths, he doesn't write Einstein's (correct) answer but, instead, adds an additional, completely unjustified factor to the equation that "turns off" the high-frequency modes and avoids the ultraviolet catastrophe. Moreover, in 1900 he does not grace his readers with any explanation for this fudge factor, which is simply introduced in the final answer.

Perhaps he felt that given his earlier statement about only using equipartition for lower frequencies, readers weren't going to take his answer seriously at high frequencies anyway. Perhaps this was his best guess about how the true law behaved at high frequencies; we will never know.[5] We do know that Rayleigh's law, with the fudge factor, was taken seriously by the German experimenters Rubens and

[5] When Rayleigh published his law with a more detailed and careful discussion in 1905, he clearly pointed out that if equipartition held for all frequencies, one would get an infinite blackbody radiation energy, which is absurd and implies some type of failure of the equipartition principle.

Kurlbaum, because they compared it to their data just a few months later. They found that while Planck's newly minted radiation law fit the data at all frequencies, Rayleigh's did so only at low frequencies, where both laws were the essentially the same.[6]

Note the different conclusions drawn by the callow patent officer Einstein and the decorated Lord Rayleigh. Einstein, through his own deep meditations on classical statistical physics, had come to the conclusion that the equipartition theorem was the only possible answer this theory could give; thus the failure of equipartition to apply to the higher frequencies of blackbody radiation meant that there needed to be a fundamentally new theory of atoms or electromagnetism, or both. Rayleigh knew that equipartition failed but hesitated to ascribe this to a failure of classical physics. In this he was similar to Planck, who was forced to accept at least the new quantum of action, h, but couldn't accept its radical implications.

But this was not the end of the story. A much younger Cambridge physicist, James Hopwood Jeans, had just completed his studies (graduating as Second Wrangler) and took up a research position, devoting himself to the theoretical study of gases. Jeans was a more flamboyant personality than Rayleigh, later specializing in astrophysics and cosmology and introducing the steady-state model of the universe, which was invalidated by the big bang theory.[7] It was precisely to blackbody radiation that Jeans turned in 1904 when he published his treatise *The Dynamical Theory of Gases*. In this work he expressed a more definitive view of the situation than Rayleigh, leading to a shocking conclusion: the equipartition theorem is valid for all frequencies, and the ultraviolet catastrophe is happening. "If an interaction between aether and matter exists, no matter how small this interaction may be . . . we are

---

[6] Editing his paper for a collection in 1902, Rayleigh added a footnote claiming that in 1900 he really meant his law to apply only to the low-frequency behavior anyway, and thus took the experiments of Rubens and Kurlbaum as vindicating his guess. The fudge factor was not mentioned. He was by then aware of Planck's correct guess and stated it.

[7] Ironically, the big bang theory was itself definitively validated by the observation of the blackbody radiation it produced.

led to the conclusion that no steady state is possible until all of the energy of the gas has been dissipated by radiation into the aether."

So why wasn't the entire earth cooled down to absolute zero as the infinity of radiation modes sucked all energy from matter? Easy. It's all happening very, very slowly.

> We can now trace the course of events when one or more masses of gas are left to themselves in the undisturbed aether [i.e., in contact with radiation] . . . a transfer of energy is taking place between the principal degrees of freedom of the molecules and the vibrations of low frequency in the aether. This . . . endows the aether with a small amount of energy. . . . After this a third transfer of energy begins to show itself, but the time required for this must be measured in millions and billions of years unless the gas is very hot.

In the technical language of thermodynamics, Jeans was dropping the assumption that matter is "in equilibrium" with radiation—the idea that radiation and matter have interacted long enough to find the most probable distribution of energy and that if one waited a long time, and measured blackbody radiation over and over again, the energy distribution would not change.

In 1905, at just about the same time that Einstein was writing his paper on quanta, Jeans and Rayleigh argued this question in a series of letters to the journal *Nature*. At this point Rayleigh, dropping his unjustified fudge factor, published what became known as the Rayleigh-Jeans law[8] in 1905: $\rho(v) = (8\pi v^2/c^3)kT$. As before, $\rho(v)$ is the mathematical

---

[8] Rayleigh published this formula with a trivial error: he neglected to include both polarizations of light, which gave too small a result by a factor of eight. Jeans corrected the error, and the law is now universally named after the pair. Einstein derived the same law independently in 1905, publishing it about a month earlier than Rayleigh, but his name is never associated with the law. Until Rayleigh's initial error was found, his formula did not agree with the low-frequency limit of Planck's law. Rayleigh noted that a comparison of the approaches would be helpful, but "not having succeeded in following Planck's reasoning, he declared himself "unable to undertake it." Once the error was corrected, the two laws agreed perfectly at low frequencies.

function expressing the radiation law—Rayleigh's classical version of the Planck law. What does it say? Well first it says that energy density of radiation at frequency v is proportional to kT. Since the equipartition theorem says that each radiation mode must have kT of energy, the factor in front, $(8\pi v^2/c^3)$, must represent the number of such modes per unit volume (hence their density). But note that this factor has a crazy feature, the one that Einstein noticed: it is proportional to the square of the frequency, implying that more and more energy is held by radiation of higher and higher frequencies. In fact the proposed law is identical to the one that Einstein wrote down and immediately rejected in his 1905 paper on light quanta, because if matter were in equilibrium with radiation the law leads to the ultraviolet catastrophe. This catastrophe is only being postponed temporarily, in the view of Jeans; the entire material universe is fighting a losing battle against radiation. The only reason we haven't all frozen to death is that we are losing the battle at an imperceptible pace.

You don't get away with this kind of maneuver scot-free. Things are interconnected in physics; in thermodynamics we almost always assume that nature is in thermal equilibrium in order to explain how things work. If you assume blackbody radiation is not in equilibrium, how, for example, do you explain the well-verified Stefan-Boltzmann law for the *total* energy radiated by a blackbody, which requires thermal equilibrium? A lucky accident?

Moreover the detailed agreement of the Planck formula with measurement would also have to be a lucky accident. And if you think hard, the coincidences required by the Jeans hypothesis multiply rapidly.[9] Planck, steeped in detailed experimental data, hardly took the Jeans idea seriously when he mentioned it in his textbook on thermodynamics in 1906. Privately he was even more scathing, commenting about

---

[9] Rayleigh, to his credit, never fully endorsed Jeans's slow catastrophe theory. Saying merely that for short wavelengths "there must be some limitation on the principle of equipartition." This limitation was provided by Planck and Einstein, but Rayleigh was not convinced, writing in 1911: "Since the date of these [1905] letters further valuable work has been done by Planck, Jeans, Lorentz, . . . Einstein and others. But I suppose the question can hardly be considered settled."

Jeans in a letter to Wien, "he is the model of the theorist as he should *not* be, just as Hegel was in philosophy; so much the worse for the facts if they don't fit." Nonetheless, the Jeans "slow catastrophe" model remained under serious discussion for the remainder of the decade.

A puzzle always looks simpler after you know the answer. It hard for us now to believe that many outstanding scientists could accept such a flimsy explanation. But, from the perspective of physicists at the time, was a radical failure of Newtonian mechanics at the atomic level really a more attractive option than dropping the assumption of thermal equilibrium? Somehow Einstein intuited that it was. And he set out to substantiate his view shortly after his acceptance of the Planck law in 1906. He did this by looking at the very same physical property that had troubled Maxwell and Rayleigh, and had led Jeans to embrace his radical alternative to the Planck law. He reexamined the specific heat of matter, but not in the gaseous state; in the solid state instead. Ultimately this work would sweep away any hope that atoms might obey classical mechanics.

CHAPTER 13

FROZEN VIBRATIONS

The whole thing started with a kind of interpolation formula by Planck. Nobody wanted to accept it because it did not appear logical . . . half the argument was continuous and the other half was based . . . on quanta of energy. The only man who appeared sensible was Einstein. He had the feeling that if there was anything to Planck's idea it must appear in other parts of physics.

—NOBEL LAUREATE PETER DEBYE, 1964

Einstein's nemesis in his student days, Professor Heinrich Weber, may have come close to depriving posterity of Einstein's historic genius, but now Herr Weber would indirectly play a crucial role in the flowering of that genius. In 1875 the young Weber, then an assistant to Helmholtz in Berlin, had just completed the best experiments extant on the specific heat of solids. The effect he was studying was an apparent violation of the empirical law noted by Pierre Dulong and Alex Petit fifty-six years earlier in 1819. These French researchers had discovered that pretty much every solid they measured had the same specific heat, once one took into account the difference of the atomic weight of the constituents. For example, a copper atom's weight is about 60 percent that of a silver atom, so 0.6 grams of copper and one gram of silver have the same number of atoms, and would then also be found to have the same specific heat. Even at that early date Dulong and Petit interpreted their finding in terms of the properties of underlying atoms, stating boldly that "one is allowed to infer . . . the following law: the atoms of all simple [elements] have exactly the same heat capacity." Later they would restrict this optimistic assessment to atoms in solids; as already noted, gases were behaving in a strange manner, which would puzzle Maxwell, Rayleigh, and others for another eighty-seven years.

What exactly is specific heat, and why did it suggest something about atoms? Specific heat is a number that characterizes a chunk of stuff (solid, liquid, gas); it is the amount the thermal energy in a gram of that stuff changes when you change its temperature by one degree centigrade.[1] Thus it measures how thermal energy varies with temperature. We have already learned that if one trusts Newtonian mechanics on the atomic scale, and the laws of statistics, then the energy of each vibrating structure bears the simplest possible relation to temperature, $E_{mol} = kT$; for atoms in a solid there are three independent directions of vibration, so according to the equipartition relation one gets $3kT$ of energy per atom.[2] If this relation holds, then if you change the temperature by one degree, the energy per atom changes by exactly $3k$, independent of the type of atom involved. This is *exactly* the law found by Dulong and Petit. But by 1906 Einstein had seen the red flag waving here; this argument relies completely on the equipartition principle, precisely the notion that he realized had failed for blackbody radiation. Thus the specific heat of solids would provide his next opportunity to extend quantum concepts, bolstered by the experimental work of his erstwhile opponent.

It is unlikely that the young Heinrich Weber would even have known the statistical theory underpinning the Dulong and Petit law when, between 1872 and 1875, he decided to test it carefully. However, earlier measurements on solids had hinted that the relation was not quite as trustworthy as its discoverers had originally thought. One elemental solid was a particularly "bad actor," one that had required "difficulty and expense" to study: diamond. Diamond, the hardest of

---

[1] Thermal energy and temperature are distinct concepts in thermodynamics. Suppose I heat a glass of water and a bathtub full of water with a blowtorch for ten seconds. Each receives the same amount of thermal energy but the change in their temperature is very different.

[2] This factor of 3 is the same one that gives us $3kT/2$ for atoms in a gas, coming from the fact that the atom can move in all three spatial dimensions.

the elemental solids, refused to give up its full quota of energy when its temperature was lowered one degree, registering a specific heat less than 30 percent of the expected Dulong-Petit value. Not only was diamond miserly with its heat energy; the measured values of its specific heat reported by various experimenters did not even agree. That is where Weber came in.

Weber, the eventual staid professor, in his youth was not averse to bold hypotheses, and so he made one: the specific heat of solids is not constant at all but can vary widely as the overall temperature is varied. This conjecture was in complete disagreement with the equipartition principle, of course, but given his distrust of theory, this would not likely have swayed Weber even had he known of it. With this hypothesis the different values of diamond's specific heat could be reconciled, as they corresponded to measurements made at rather different starting temperatures. Weber suspected that somehow the Dulong-Petit value of $3k$ per atom was only reached at high enough temperatures, and for some reason, in the case of diamond, room temperature wasn't high enough. He thought that if he could cool diamond samples well below room temperature, he would find even larger deviations from that value. His work predated all the breakthroughs in cryogenics that now make it possible routinely to lower the temperature of a solid to hundredths of a degree above absolute zero ($-273°C$). Poor Weber had to rely on natural ice to do his measurements at low temperature, and needed to suspend them in March of 1872 due to the lack of available snow!

By 1875 Weber had pushed his experimental technology to a higher level and was able to present beautiful measurements of the specific heat of diamond, varying the temperature from $-100°C$ to $+1,000°C$. Sure enough, at the highest temperatures the specific heat of diamond increased until it attained the Dulong-Petit (DP) value and then stopped increasing, whereas as the temperature was lowered below normal room temperature it continued to decrease down to onefifteenth of the DP value. Moreover other elemental solids showed a similar but less dramatic variation with temperature. Weber's basic hypothesis was right. For some reason, for most materials, room temperature *is* hot enough that the DP law initially appeared universal; but for diamond and a few others it is not. And most puzzling of all, at very low temperatures diamond and other materials appeared to lose completely the ability to emit or absorb heat energy when the temperature was changed; their specific heat seemed to disappear. Walther Nernst, who studied with Weber prior to becoming the preeminent physical chemist of his generation, described the situation thus: "through the diamond experiment one has therefore found that the atomic vibrations can be brought to a standstill. As soon as this happens, the concept of heat does not any longer exist for the 'dead body.'" Weber had made a great experimental discovery, the greatest of his career; eventually it landed him a full professorship at the Poly, leading to his fateful encounters with Einstein.

Recall that Einstein lauded Weber's course on heat during their brief honeymoon period. Einstein's course notes from that time have actually survived, but they contain no evidence that Weber discussed his own discovery, the strong temperature variation of specific heat. Nonetheless we have already noted that by 1901 Einstein, in a letter to Mileva, announced that he had been considering "the latent [specific] heat of solids" in connection with Planck's radiation formula, and that his views on latent heat had changed *because* his views on radiation theory had "sunk back into the sea of haziness." Thus it is safe to assume he was by then aware of Weber's systematic demonstration of anomalous behavior. Now, in early 1906, Einstein's views on radiation were no longer hazy: Planck's formula was right, equipartition was wrong, and Newtonian mechanics was in jeopardy. It was time to see if the heretical ideas relating to quanta could clean up the specific heat anomalies just as they had explained the odd behavior of the photoelectric effect. By November of 1906, eight months after his paper announcing that the Planck formula *required* light quanta, Einstein submitted his second great work on quantum theory to *Annalen Der Physik*, titled "Planck's Theory of Radiation and the Theory of Specific Heat."

Einstein's papers in general have a more philosophical tone than typical physics papers, even those of the time. And so after an introductory review of his 1905 and 1906 papers on light quanta, he presents the following ontological dictum to the (in all likelihood dumbfounded) reader:

For although one has thought before that the motion of molecules obeys the same laws that hold for the motion of bodies in our world of sense perception . . . we must now assume . . . that the diversity of states that they can assume is less than for bodies within our experience. For we make the additional assumption that the mechanism of energy transfer is such that the energy of elementary structures can only assume the values 0, $h\upsilon$, $2h\upsilon$, etc.

This is the statement of quantization of energy at the atomic scale, as clear and unequivocal as one would find in a modern physics textbook. Einstein, not Planck, said it first. Discontinuity is not a mathematical trick; it is the way of the atomic world. Get used to it.

Einstein continues:

I believe we must not content ourselves with this result. For the question arises: If the elementary structures . . . cannot be perceived in terms of the current molecular-kinetic theory [of heat], are we then not obliged also to modify the theory for other periodically oscillating structures considered in the molecular theory of heat? In my opinion the answer is not in doubt. If Planck's radiation theory goes to the root of the matter, then contradictions between the current molecular-kinetic theory and experience must be expected in other areas of the theory of heat as well, which can be resolved along the lines indicated. In my opinion this is actually the case, as I now shall attempt to show.

The argument from here is remarkably straightforward. Atoms form a solid when they arrange themselves in a regular pattern in space, held together by electrostatic interactions. Einstein states that the simplest picture one may have of heat energy stored in a solid is that all the atoms "perform [periodic] oscillations around their equilibrium positions." As already noted, for a mass oscillating periodically back and forth in each of three directions the equipartition principle predicts $3kT$ of energy per atom, yielding the DP value for the specific heat. But, Einstein notes, several elements (diamond, boron, silicon) have smaller specific heat than expected from this law, and compounds containing oxygen and hydrogen also show similar violations. Finally, he notes that Drude identified other kinds of oscillations in solids, involving the electrons, which appear to be important in how solids absorb light but don't seem to contribute to the specific heat. But the equipartition principle requires that *all* oscillations get their share of energy, so these "extra" oscillations should cause the specific heat of solids to actually exceed the DP value, which was not observed. So something is out of kilter.

The atoms in a solid were really no different from the "elementary resonators" in Planck's blackbody radiation theory (which were held in place by electric forces but could vibrate in all three directions around their equilibrium values), and Einstein had *already* announced in his 1906 paper that such vibrating structures can only have energies equal to an integer times their frequencies, $E = 0$, hv, $2hv$, etcetera. Thus each atomic vibration has a ladder of allowed energies separated by hv. But the typical amount of energy available to each atom from its thermal environment is just the equipartition value, kT (per direction of vibration). So what happens if the quantized energy of the atomic vibration, hv, is much larger than kT? The atom then is like a man trying to climb a ladder whose rungs are much farther apart than his reach. It can never get off the lowest "rung"; its vibrational energy remains stuck at zero.

Thus some modes of vibration are "frozen out"; their first nonzero quantized energy level is too high to absorb the amount of energy dictated by the Dulong-Petit (equipartition) law. Moreover, it makes sense that these "missing vibrations" would disappear first in materials that are very hard, like diamond. Roughly speaking, a material is harder if its atomic constituents are more tightly bound in place, so that they vibrate very rapidly when disturbed from equilibrium. But if they vibrate very rapidly, then their frequency is unusually high, so that the energy-level spacing of that material, hv, is unusually large. Thus, when compared over the same range of (decreasing) temperature, their vibrations freeze into the lowest level before those of a softer solid. This paucity of high-frequency atomic vibrations is of course conceptually linked to the "missing" high-frequency modes of thermal radiation that characterizes the Planck law and that so puzzled Rayleigh. Einstein had now realized that quantum freezing of vibrations is also the ultimate explanation for the strange behavior of the specific heat of solids.

However, to actually get a precise formula for the quantum specific heat of a solid that he could compare to data, Einstein decided to make a simplified model of a vibrating solid. Any system in mechanical equilibrium will oscillate back and forth when it is given a little energy; think of a pendulum pushed a bit to the side from the vertical. But the frequency of the oscillations depends on the details of the system, and for a solid made up of an enormous number of atoms there are many different types of oscillatory motions with many different frequencies, depending, for example, on the chemical bonding arrangements of the constituent atoms. This set of different frequencies was too complicated to work out at the time (modern quantum physicists can do it with incredible precision), so to compare his theory's prediction to the measurements for diamond, Einstein assumes that it has only a single, primary frequency of vibration. He is quick to point out that, given this simplification, "of course an exact agreement with the facts is out of the question."

Nonetheless, this assumption gives him an approximate law for the temperature variation of specific heat based directly on Planck's expression for the energy of a single oscillator of frequency v; and this expression shows remarkably good agreement with Weber's data. He remarks, "both above-mentioned difficulties[3] are resolved by the new interpretation and I believe it likely that the latter will prove its validity in principle." In fact, according to Einstein's new theory, the specific heat of *all* solids decreases with decreasing temperature until, at the absolute zero of temperature, it completely disappears—a stunning prediction.

But there was one further radical step to take. Throughout this paper Einstein assumes that the same molecular vibrations that store heat also exchange energy with radiation through emission and absorption, thus closely tying the specific heat formula to the blackbody

---

[3] The disappearance of specific heat at low temperature, and the absence of "extra" specific heat due to optical-frequency electronic vibrations.

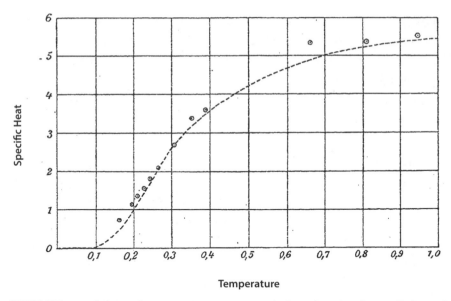

FIGURE 13.1. Graph from Albert Einstein's 1907 paper which predicts that the specific heat of all solids should go to zero as the temperature is lowered, due to quantization of vibrational energy. Here the theory (dashed line) is being compared to Weber's data for the temperature variation of the specific heat of diamond. Courtesy the Albert Einstein Archive.

law. But after submitting his paper he recalled that there are molecular vibrations that do not interact with radiation at all, and that such vibrations can still store heat and contribute to the specific heat.[4] Einstein realized that this was an important observation and actually published a note of correction, stating, "most certainly there could exist uncharged heat carriers [vibrations], i.e. such ones that are not observable optically." But if neutral vibrations, those that do not interact with radiation, were also subject to the law of quantization of energy, then whatever the quantum theory was, its domain was not merely the interaction of radiation with matter, as Planck had hoped. The disease of discontinuity was present in matter without radiation; Newtonian atoms had frozen to death.

[4] These are vibrations that do not generate a net dipole moment.

CHAPTER 14

PLANCK'S NOBEL NIGHTMARE

The two constants [h, k] . . . which occur in the equation for radiative entropy offer the possibility of establishing a system of units for length, mass, time and temperature which are independent of specific bodies or materials and which necessarily maintain their meaning for all time and for all civilizations, even those which are extraterrestrial and non-human.
—MAX PLANCK

It was the fall of 1908, and Svante Augustus Arrhenius was determined to see that Max Planck received the Nobel Prize for Physics that year. Arrhenius, a scientist of impressively broad and bold speculations, had recently returned from a tour of Europe, where he was received warmly as befitted the first Swedish winner of the newly minted Nobel prizes. Arrhenius had won the Chemistry Prize in 1903 (two years after the establishment of the awards) for his groundbreaking work on electrolytic chemistry. He was widely recognized as a founder of the discipline of physical chemistry, which works at the boundary of the fields of physics and chemistry. In 1905 he had been offered a professorship in Berlin but had turned it down to remain in Sweden and head the new Nobel Institute for Physical Research; after receiving the prize he would be a member of the Nobel Award Committee in Physics and a de facto member of the Chemistry Committee for the remainder of his life. As such he had enormous influence over who received these awards, and he did not hesitate to use that influence.

Arrhenius, like all his contemporaries, was blissfully unaware of the looming crisis in atomic physics, uncovered by the work of the young Einstein, who was now becoming known—not for challenging the Newtonian paradigm of continuous motion but instead for dismissing another Newtonian axiom, the concept of absolute time. While Einstein had quickly moved to the terra incognita of the nascent quantum theory, *assuming* that atoms existed and trying to figure out their laws of motion and their interactions with radiation, Arrhenius was still fighting the last war, the war to prove that atoms were real. The ensuing episode illustrated just how oblivious the scientific community was to the gathering storm.

Had Arrhenius known the story of the checkered career of the German/Swiss Jew, who was still not recognized formally by the conservative professoriate of Switzerland in 1908, he likely would have recognized a kindred spirit. Arrhenius grew up near Uppsala, Sweden, where his father was a surveyor for the University of Uppsala, the oldest and among the most prestigious of the Nordic universities. A science and math prodigy, he had matriculated at the university at age seventeen, and received his degree in two years, before moving on to graduate studies in physics. However, in a striking parallel to Einstein, he alienated the senior members of the faculty, Tobias Thalen (physics) and Per Theodor Cleve (chemistry), and left after three years to complete his doctorate at the new Physical Institute of the Academy of Sciences in Stockholm. Unfortunately for Arrhenius the new institute was not yet allowed to grant PhDs on its own. Thus when, in 1884, he produced a monumental 150-page work on the conductivity of electrolytic solutions, explaining, for example, the high conductivity of salt in water by its dissociation into ions, it was received with great skepticism by a committee consisting mainly of faculty whom he had spurned at Uppsala. In the end the thesis was approved with the lowest possible passing grade, *non sine laude approbatur*, ("accepted, not without praise"). Forty years later Arrhenius would recount bitterly that Cleve and Thalen even refused to offer him the customary congratulations after the doctoral ceremony, saying that they had decided to "sacrifice him." Although this work and its extensions would eventually earn him the Nobel Prize, the grade it had received was so poor that he was at least nominally disqualified from pursuing an academic career in Sweden at the time.

Here, however, his story diverges from that of Einstein, for he boldly sent the devalued thesis to the leading lights of European chemistry and physics, Clausius (inventor of the concept of entropy), van 't Hoff in Amsterdam (who would be the first Nobel Laureate in Chemistry), and Ostwald in Riga (the ninth Nobel chemistry laureate). One of these men, Ostwald, immediately recognized its innovativeness, to the extent that he even traveled personally to Uppsala to offer Arrhenius a job at his own institution.[1] Arrhenius did not cut a particularly impressive figure, according to Ostwald: "[Arrhenius] is somewhat corpulent with a red face and a short mustache, short hair; he reminds me more of a student of agriculture than a theoretical chemist with unusual ideas." But a brilliant chemist he was, and eventually Arrhenius did move to Europe and trained with Ostwald, van 't Hoff, and even with Boltzmann before returning to Sweden to become the unquestioned leader of Swedish physical chemistry, and the person who defined the international scope of the Nobel prizes at their inception.

A decade later, at the turn of the century, there was still a major movement in chemistry and physics that regarded atoms as somewhat suspect heuristic entities, a movement led by Arrhenius's former mentor, Wilhelm Ostwald. This school of thought was known as "energetics" and also had adherents in the Swedish physics community, which maintained an attitude of distrust toward theory in general and of "pronounced hostility toward atomism and toward atomic theory" in particular. Arrhenius had decided to put this movement to final rest and make 1908 the Nobel Year of the Atom. Max Planck would receive the physics prize for the manner in which his radiation law had led to an accurate determination of Avogadro's number and

[1] There is some irony here in comparison with Einstein, who sixteen years later would write to the very same man asking for a job, leading to a famous letter from his father to Ostwald (behind Einstein's back) essentially pleading with Ostwald to assent. No known answer was received, but in 1910 Ostwald became the first scientist to nominate Einstein for the Nobel Prize.

the elementary unit of atomic charge, e. The chemistry prize would be awarded to the British physicist Ernest Rutherford, who had shown that atoms disintegrated (i.e., emitted doubly ionized helium atoms, known as alpha particles) during radioactive decay. In a very recent experiment with Geiger, Rutherford had deduced a value of the elementary charge from alpha particles in excellent agreement with that calculated by Planck using his radiation law, tying the two prizes neatly together.

The fact that Rutherford considered himself a physicist and would be very surprised to know that he had been reclassified a chemist[2] did not deter Arrhenius from his plan. Arrhenius had nominated Rutherford for *both* the physics and chemistry prizes that year, but it is likely that he had planned all along to support Planck in the Physics Committee, of which he was a member. By the time of the crucial meeting on September 18, 1908, he knew that the Chemistry Committee (based on an internal report he had apparently ghostwritten) was committed to awarding the prize to Rutherford. Planck and Wien had been jointly nominated in physics for the theory of heat radiation by Ivar Fredholm, a Swedish mathematician and mathematical physicist, and Arrhenius swung his support to this nomination, but with the intention of splitting the ticket and engineering a prize for Planck alone.

Why did Arrhenius think that Planck alone should be recognized? Because at that time Arrhenius was not interested in the physical principles behind the law of thermal radiation[3] so much as in its connection to the fundamental constants in molecular chemistry. This is an aspect of Planck's work of 1900 that is barely mentioned in modern times, but at that time it overshadowed his radical quantum hypothe-

---

[2] Rutherford, who had spent many years proving that elements transmuted during radioactive decay, later joked that "he had seen many transmutations in his time but none as quick as his own transmutation into a chemist."

[3] Arrhenius did have another reason to be interested in thermal radiation. He was the first scientist to recognize the role of CO_2 in trapping heat radiation and warming the planet, and even suggested the possibility that human-generated industrial CO_2 emissions would enhance this effect. He published this idea in that same year, 1908, arguing that it was a good thing and might forestall future ice ages.

sis. Planck's radiation law depended on the two newly discovered physical constants that he introduced, h, the "quantum of action" (Planck's constant), and k, Boltzmann's constant (the constant associated with entropy through the equation $S = k \log W$ and thermal energy through the equipartition relation $E_{mol} = kT$.) From a careful fit of blackbody radiation data one can extract quite precise values for both h and k, and Planck did so immediately after deriving his radiation law in 1900. The constant h appeared to him completely enigmatic and was not put to any immediate use, but the constant k, which only later became known as Boltzmann's constant,[4] was instantly recognized as providing a theoretical microscope for studying the atom.

In his December 1900 magnum opus Planck states, "To conclude, I may point to an important consequence of this theory which at the same time makes possible a further test of its reliability." He goes on to show by straightforward steps that the Boltzmann constant satisfies the simple relationship $k = R/N_a$, where R is the constant in the ideal gas law $PV = RT$ for a mole of gas, and N_a is Avogadro's number (which has struck fear into so many beginning chemistry students), the number of atoms contained in a mole of any gas. This number was imperfectly known in 1900, whereas R was very well known. Hence by extracting k very precisely from the radiation law, Avogadro's number could be determined to unprecedented precision. Planck found the value $N_a = 6.175 \times 10^{23}$, which is within 2.5 percent of the currently accepted value 6.022×10^{23}. Using the same information, he could determine the mass of a hydrogen atom, again with high accuracy. Finally, in a coup that must have impressed the physical chemist, he used considerations from electrolytic chemistry, Arrhenius's own field, to find the elementary charge on a proton, obtaining a value within 2.5 percent of the modern value. In contrast the best-known value of e, the charge on an electron, measured by J. J. Thomson from electron studies, was off by 35 percent! Planck concluded his 1900 analysis with the confident declaration, "If the theory is at all correct, all of these

---

[4] It was initially called by many "Planck's constant," leading to no end of confusion until the conventions settled down, assigning h to Planck and k to Boltzmann.

relations should not be approximately, but absolutely valid. The accuracy of the calculations . . . is thus much better than all determinations up to now."

Planck had always been fascinated by fundamental constants as expressions of the absolute and eternal in physics. Even before his work of 1900 he had realized that the radiation law involved two distinct and new fundamental constants. Fundamental constants allow one to define what are called *absolute units*, units of measurement relating to the basic laws of physics. For example, the speed of light, c, provides a natural unit of velocity, because no signal can travel faster than c and all relativistic phenomena become more and more important as this speed is approached. In the famous twin paradox of special relativity, your identical twin ages more and more slowly compared with you as her relative velocity approaches c. Planck pointed out that his two newly discovered constants, when combined with the speed of light and the gravitational constant, would allow fundamental units to be defined for *all* physical quantities (length, time, temperature, etc.). Transported by this revelation, the staid professor allowed his inner geek to emerge in print, rhapsodizing that these units would be valid for "all times and civilizations . . . even extraterrestrial ones." Later, when Planck became embroiled in a philosophical debate with the Viennese philosopher-scientist Ernst Mach, Mach would lampoon his exuberance over fundamental units: "concern for a physics valid for all times and all peoples, including Martians, seems to me very premature and even almost comic."

Nonetheless in 1900 it was these fundamental constants, which had emerged from his radiation law, that most excited Planck, and not his unexamined introduction of discontinuity into the laws of physics. To his disappointment, the rest of the physics community did not immediately appreciate even this aspect of his breakthrough. He later recounted:

I could derive some satisfaction from these results. But matters were viewed quite differently by other physicists. Such a calculation of an elementary electrical [charge] from measurements of thermal radiation was not even given serious consideration in some quarters. But I did not allow myself to become disturbed by such a lack of confidence in my constant k. Nevertheless, I only became completely certain on learning that Ernest Rutherford had obtained a [very similar] value by counting alpha particles.

This spectacular agreement between completely disparate physical phenomena, all pointing to a single consistent atomic picture of the world, had convinced Arrhenius that Planck alone should be recognized with the physics Nobel Prize in 1908. It was the connection to fundamental constants that distinguished Planck's work from Wien's in Arrhenius's mind, and in his report to the Nobel Physics Committee he barely mentioned Planck's derivation of the radiation law and completely omitted any mention of "quanta of energy." Planck's use of the constant k, he said, had "made it extremely plausible that the view that matter consists of molecules and atoms is correct. . . . No doubt this is the most important offspring of Planck's magnificent work."

Arrhenius's enthusiasm did not sweep through the conservative Physics Committee unchallenged. Among its members was the distinguished experimentalist Knut Angström, who had actually done experiments on heat radiation and was aware of the experimental prehistory leading up to Planck's "act of desperation." With much justice he wrote, "it is very far from being that the theoretical works have guided the experimental ones, but rather that one could justly make a completely contrary statement." However, there was a small problem with his argument that an experimenter should receive or share the prize: none had been nominated that year. Angström and the other skeptics on the Physics Committee were reluctantly convinced by Arrhenius to join the Planck bandwagon.

And so the modest, upright Planck (who had himself nominated Rutherford for the physics prize that year) might have received this honor, not because of a deep appreciation of the true significance of his work, elucidated by Einstein from 1905 to 1907, but rather because of a general *ignorance* of its full implications. After the Physics Section of the Swedish Academy had approved Planck as the awardee, rumors of the result quickly traveled around the continent, apparently reaching Planck himself, who stated to the press, "[if true] I presume that I owe this honor principally to my works in the area of heat radiation." But the full Swedish Academy would still have to approve the recommendation of the Physics Committee, and in the interim between these votes something had changed the mood in Stockholm. The most famous theoretical physicist of his generation, the man Einstein admired the most, had finally spoken publicly on the Planck law, and his opinion would derail Arrhenius's well-laid plans.

Hendrik Antoon Lorentz was born on July 18, 1853, in Arnhem, the Netherlands, to an unexceptional middle-class family. His extraordinary brilliance was recognized early, and by the age of twenty-four he was appointed to the newly created Chair of Theoretical Physics at the University of Leiden. He devoted his early years to the application and extension of Maxwell's theory of radiation. In particular, while J. J. Thomson is credited with "discovering the electron" in 1897, Lorentz *deduced* its existence a year earlier, in 1896, from his analysis of light emitted from a gas in the presence of a magnetic field—the "Zeeman effect," discovered by his former student and assistant Pieter Zeeman. He shared the Nobel Prize in Physics with Zeeman in 1902 for this work (the first theorist to be so honored) and went on to develop an elegant theory of the interaction of electrons with light, published in 1904. In related work, Lorentz came to the very edge of the special theory of relativity, coming up short only by his unwillingness to interpret relativistic effects as arising from the relative nature of time, as did Einstein in 1905. In fact Lorentz was troubled by Einstein's approach, complaining, "Einstein simply postulates what we have deduced with some difficulty and not altogether satisfactorily, from the fundamental equations of the electromagnetic field." Despite these misgivings, within a few years Lorentz became Einstein's close confidant and scientific father figure, supporting and providing constructive criticism for all his major research.

We have already heard that Einstein regarded Lorentz as the most powerful thinker he had ever encountered, but his admiration ran deeper than this because of Lorentz's elegant, generous, and kindly spirit. Late in his life Einstein wrote: "Everything which emanated from his supremely great mind was as clear and beautiful as a good work of art. . . . For me personally he meant more than all of the others I have met in my life's journey. Just as he mastered physics and mathematical structures, so he mastered also himself—with ease and perfect serenity." Einstein's attitude toward Lorentz was, by 1908, shared by much of the European physics and mathematics community. His encyclopedic knowledge of many subdisciplines made his opinion the final word for many. "Whatever was accepted by Lorentz was accepted; and whatever was rejected by him was rejected or at best labeled controversial." He, alone among all theorists, had preceded Einstein in noting Planck's essential use of the "energy element" in his derivation of the radiation law as early as 1903, and had "wrestled continuously with this problem [of heat radiation]" in the years leading up to 1908.

Lorentz, with his unmatched mastery of both electromagnetic theory and the dynamics of electrons, set out to *deduce* the Planck radiation law rigorously, directly from the motion of charged electrons, which radiate and absorb radiation, without introducing fictitious molecules or "resonators" as Planck had done. But no matter how hard he tried, he succeeded only in rediscovering the low-frequency, approximate law of Rayleigh-Jeans (and Einstein), which led to absurd consequences (infinite energy) at high frequencies. In April of 1908 he announced his findings at the International Congress of Mathematicians in Rome.

> The theory of Planck is the only one that would provide us with a formula in accord with experimental results; but we could accept it only with the stipulation that we completely rework our basic conception of electromagnetic phenomena. . . . However I must address myself to the question of how the Jeans theory, which involves no constants other than . . . [Boltzmann's constant] k, can take into account the peak of the radiation curve which has been demonstrated by experiments [the absence of high-frequency thermal radiation predicted by Jeans]. The explanation given by Jeans—which is really the only one that can be given—is that the maximum is illusory; its existence is simply an indication that it has not been possible to realize a body that is black to short wavelengths. . . .

Fortunately, we can hope that new experimental determinations of the radiation function will permit a choice between the two theories.

Lorentz, the most respected theoretical physicist of his generation, was leaning in favor of the Jeans "slow catastrophe" theory!

The German physics community was initially stunned, and then apoplectic. Two outstanding experimenters, Lummer and Pringsheim (whom Ångström had favored for the 1908 Nobel Prize) wrote scathingly, "If we examine the Jeans-Lorentz formula, we see at first glance that it leads to completely impossible consequences which are in crass conflict not only with the results of all observations of radiation, but with everyday experience." They continued with more than a hint of sarcasm: "We might therefore dismiss this formula without further examination were it not for the eminence and authority of the two theoretical physicists who defend it." Wilhelm Wien expressed the outrage felt by many when he stated, "I was extremely disappointed by the lecture which Lorentz presented in Rome. . . . he came up with nothing more than the old theory of Jeans without adding any new viewpoint. . . . [That theory] is not worthy of discussion in the experimental field. . . . What purpose is served by submitting these questions to mathematicians, since they can provide no judgment in this matter? . . . *In this instance Lorentz has not shown himself to be a leader of science*" (italics added).

Subjected to this torrent of criticism, even the serene Lorentz must have had second thoughts. By June 1908, only two months later, he wrote a long letter to Wien, which contained an apology in the form of an embarrassing thought experiment. In the Rayleigh-Jeans theory (which Lorentz had rederived), the energy emitted as heat radiation at a given frequency is proportional to the temperature. A metal such as silver heated to 1200°K (roughly 900°C) will glow with blinding white light. When the temperature is reduced to room temperature (roughly 300°K), it should still be emitting one-quarter as much white light, according this theory. Hence, Lorentz continued, if the theory were right an unheated silver mirror would glow visibly in the dark! He concluded, "thus we should really dismiss the Jeans theory . . . and we are left with only the theory of Planck. Do not think that I do not respect it, quite the contrary, I admire it greatly for its boldness and success."

Lorentz's speedy retraction and his professed admiration for Planck's theory, expressed privately, did not get as much publicity as his public criticism of it in Rome. A mathematician with substantial influence in the Swedish Academy, Gosta Mittag-Leffler, had gotten wind of Lorentz's critique and used it to his advantage. He wished to ingratiate himself with the French clique, led by the great mathematician Poincaré, and to divert the physics prize to the second-choice candidate, the French pioneer of color photography Gabriel Lippmann. In this scheme he was aided by the original nominator of Planck and Wien, the mathematician Ivar Fredholm, who was unhappy with the elimination of Wien from the award. Fredholm wrote to Mittag-Leffler before the full academy vote was taken, criticizing the Physics Committee decision and stating specifically that Planck had based the derivation of his radiation law on "a completely new hypothesis, which can hardly be considered plausible, namely the hypothesis of the elementary quanta of energy."

With the sudden realization that Planck's law was radical and controversial, his nomination was soundly defeated in the general academy vote, and the prize was instead awarded to Lippmann. Mittag-Lefler sent a gleeful letter taking credit to a French mathematician: "It is I, along with Phragmen, who got the prize awarded to Lippmann. Arrhenius wanted to give it to Planck in Berlin, but his report, which he had somehow gotten accepted unanimously by the Committee, was so stupid that I was able to crush it easily." The physics world was only beginning to grapple with the inevitability of the quantum revolution, proclaimed by the Valiant Swabian in his great works of 1905–7. Planck, Lorentz, and the other great scientists of Europe would have to look to this fearless interloper to lead them forward in the quest for a new microscopic mechanics. More than a decade would pass, and much would happen, before the Nobel committee would again be willing to consider giving the prize for an atomic theory.

CHAPTER 15

JOINING THE UNION

"So, now I too am an official member of the guild of whores." Thus Einstein announced to a friend, Jakob Laub, his long-overdue acceptance into the Swiss professoriate. Einstein had been appointed extraordinary professor of theoretical physics at the University of Zurich on May 7, 1909, and was writing shortly thereafter to describe the last phase of his "hazing" before admission to the fraternity. This same Laub, a young Austrian physicist who had studied with Wien, had written Einstein fourteen months earlier, saying, "I must tell you quite frankly that I was surprised to read that you must sit in an office for eight hours a day. History is full of bad jokes." Laub was reflecting the general amazement in German physics at the mysterious oracle who had emerged in Switzerland with no fanfare, the protégé of no great man, and appeared to be rewriting physical theory as a hobby between reviewing patents. (There may be some truth to the latter in that Einstein admitted to hiding his physics calculations in his desk at the patent office for occasional perusal.)

What Einstein had done during his first five years while working six days a week at the patent office was beyond astonishing. From 1902 to 1904 he produced his fundamental studies in statistical mechanics, which were underappreciated but laid the foundations for many of his later breakthroughs. In 1905, the miracle year: light quanta, Brownian motion, special relativity, and $E = mc^2$, all eternal contributions to the physics canon. In 1906: the quantum theory of specific heats and the announcement that a quantum mechanics was needed for any correct atomic theory. In 1907: a masterful review article on relativity theory, which contained for the first time the "happiest thought" of Einstein's life, the thought that a uniform acceleration was indistinguishable from the presence of a uniform gravitational field. This idea became known as the equivalence principle and provided the germ of the general theory of relativity, Einstein's magnum opus. Imagine what the guy could have done if he had actually had some time to focus on physics.

Having made these modest contributions to science, Einstein understandably began to think that he might now find employment in a university setting, where research would actually be part of the job description. But the Swiss physics establishment was not easily diverted from its traditional habits. The University of Bern had a stodgy, backward department, run by "a few old fogies"; Einstein had dismissed it as a "pigsty" soon after moving to Bern. To become a teacher at a university in the Germanic countries, one had to submit a so-called habilitation thesis, a more extensive and original work beyond the doctoral thesis. Einstein submitted a group of seventeen of his research papers for habilitation at the University of Bern in June of 1907, but the application was rejected, ostensibly because he had not integrated them into a single handwritten thesis, but there were other factors at work as well. The professor for experimental physics, Aimé Forster, saw no real value in adding a theorist to their faculty, and allegedly returned Einstein's paper on special relativity with the comment, "I can't understand a word of what you have written here." Finally, in February 1908, he provided a more acceptable thesis, not a compilation of his papers but a specific work on the blackbody law and the constitution of radiation,[1] and on this basis Einstein was granted the status of *Privatdozent* (private instructor) in physics at the university. This effort entitled him to provide additional lessons for the physics students, with no salary, on top of his full-time work at the patent office. However it was a necessary step toward obtaining a professor's position.

A few months later the pressure began to mount to end the embarrassment to Swiss science and award Einstein an actual academic position in theoretical physics. Lorentz and Minkowski spoke glowingly

---

[1] Sadly, this thesis, which would have provided a window into his thinking at a fascinating moment, has been lost.

of him at the infamous Rome Congress, at which Lorentz unwittingly undermined Planck's Nobel nomination; and Planck's great respect for Einstein's work was widely known. It so happened that an appropriate position *was* being created at the University of Zurich, *Professor Extraordinarius*, to lighten the load of his nominal thesis adviser, Alfred Kleiner, the *Professor Ordinarius*. Despite its impressive sound in translation, an extraordinary professor was actually a large notch below the "ordinary professor"; essentially he was a subordinate of Kleiner, with lower salary and fewer perks. Nonetheless, it was a real portal into an academic career for Einstein, and an opportunity for Swiss physics to improve its standing on the continent.

But parochialism would not die easily; Kleiner favored a local scholar from a well-known family, Friedrich Adler, for the position. After some back-and-forth, Adler, who was not an outstanding physicist and really was more interested in philosophy, withdrew his name from consideration, writing to his father, "[Einstein] will most likely get the professorship—a man who on principle . . . should certainly get it rather than myself . . . and if he gets it I will . . . be very pleased. . . . [He] was a student at the same time as I, . . . the people involved . . . have a bad conscience about the way they treated him in the past, and . . . it is felt to be a scandal, not only here but also in Germany, that a man like that should sit in the Patent Office. . . . Objectively . . . it is a fine thing that this man has asserted himself despite all difficulties."

Even *with* Adler's withdrawal, Einstein was not immediately offered the job, as Kleiner deemed his teaching ability inadequate. Einstein had to ask Kleiner for a second teaching evaluation, through an arranged lecture at the Zurich Physics Society, and this time was successful, as he explained to Laub: "I was really lucky. Totally against my usual habit I lectured well." In a final indignity Einstein was offered a considerably smaller salary than he was already earning at the patent office, but when he refused to take the position under those circumstances, it was agreed his patent salary would be matched. The formalities were then executed successfully, and Einstein was inducted into the guild. It is no wonder that Einstein was not feeling too grateful for his new position when it finally materialized in May of 1909.

Einstein's friend Laub, to whom he announced his appointment, had previously come to visit Einstein and work with him while he was still in Bern, shortly after sending his 1908 letter terming Einstein's situation a "bad joke." He thus became the first scientist to actually coauthor a paper with Einstein (prior to this Einstein had been the sole author of every one of his works). In fact, although Einstein would collaborate intermittently through his career and he greatly enjoyed discussing physics with colleagues, none of his greatest papers were to have coauthors,[2] and his collaboration with Laub was no exception. They produced together two undistinguished papers on relativity theory, and soon after Laub departed in May of 1908, Einstein set relativity theory aside and devoted himself solely to radiation/quantum theory. During the summer and fall of 1908, while Lorentz was still vacillating between the Rayleigh-Jeans and Planck laws, and the kingpins were wrangling over Nobel prizes in Stockholm, Einstein had digested the profound implications of quantization of energy and was trying to make sense of light quanta.

Already, in January of 1908, Einstein had indicated that his principal concern was not relativity theory but understanding the quantum theory, which he now had shown comprised both radiation (through light quanta) and atomic mechanics (through the quantum law of specific heat). At that time he replied to an inquiry from Arnold Sommerfeld, the other great German theorist of the time (along with Planck). Sommerfeld had succeeded to Boltzmann's chair in theoretical physics in Munich in 1905. He was Prussian through and through, with a large mustache and a bearing that gave the "impression of a colonel of the Hussars," accentuated by a dueling scar on his face acquired as a member of the drinking and fencing society (the *Burschenschaft*) during his student days. Not only was his appearance

---

[2] Much later in his career (after 1925) Einstein consistently worked with collaborators but failed to produce papers of historic significance, with one exception. In 1935 he published with Boris Podolsky and Nathan Rosen a profound work, intended as a critique, pointing out the nonlocal nature of quantum mechanics ("the EPR paradox"). In fact this feature of the theory has been experimentally confirmed and plays a key role in the important and growing field of quantum information physics.

intimidating; so also was his intellect, reflecting a mathematical facility rare even among theoretical physicists. His initial reaction to Einstein's work hints at some resistance to a new Jewish "prophet"; in a letter to Lorentz, Sommerfeld wrote: "we are now all longing for you to comment on that whole complex of Einstein's treatises. Works of genius though they are, this unconstruable and unvisualizable dogmatism seems to me to contain something almost unhealthy . . . perhaps it reflects . . . the abstract-conceptual character of the Semite." This initial bias dissipated rapidly, however, and soon he would speak of Einstein with great respect and make major contributions to both relativity and quantum theory.

In his January letter replying to Sommerfeld, Einstein seems a bit embarrassed by a previous complimentary letter from him (which is lost) and begins thus: "Your letter made me uncommonly happy. . . . Thanks to my having hit upon the fortunate idea of introducing the relativity principle into physics, you (and others) enormously overestimate my scientific abilities . . . let me assure you that if I were in Munich . . . I would sit in on your lectures in order to perfect my knowledge of mathematical physics." Then Einstein gets to the point. In response to a query from Sommerfeld, he states that relativity theory does not at all provide a definitive theory of electrons: "a physical theory can be satisfactory only when it builds up its structures from *elementary* foundations." Relativity theory, he says, is like thermodynamics *before* Boltzmann explained what entropy means at the atomic level.

> I believe that we are still far from having satisfactory elementary foundations for electrical and mechanical processes. I have come to this pessimistic view mainly as a result of endless, vain efforts to interpret the second universal constant [h] in Planck's radiation law in an intuitive way. *I even seriously doubt that it will be possible to maintain the general validity of Maxwell's equations for empty space* [italics added].

A year later, in January 1909, Einstein put this radical idea into print in a paper titled, "On the Present Status of the Radiation Problem," which followed papers of similar focus from Lorentz, Jeans, and

Walter Ritz.[3] After some preliminaries he again states the fundamental contradiction they were all facing, "There can be no doubt . . . that our current theoretical views inevitably lead to the law propounded by Mr. Jeans. However we can consider it . . . equally well established that this formula is not compatible with the facts. Why, after all, do solids emit light only above a fixed, rather sharply defined temperature? Why are ultraviolet rays not swarming everywhere . . . ?" After showing again how quantization of energy leads to Planck's law, he states, "Though every physicist must rejoice that Mr. Planck disregarded [the requirements of classical statistical mechanics] in such a fortunate manner, it should not be forgotten that the Planck radiation formula is incompatible with the theoretical foundation from which Mr. Planck started out."[4]

Einstein then presents a new and subtle argument (which will be described below) for why the Planck formula tells us that light has both particulate and wave properties simultaneously. He concludes, "In my opinion, the last . . . considerations conclusively show that the constitution of radiation must be different from what we currently believe"; therefore "the fundamental [Maxwell] equation of optics will have to be replaced by an equation in which the [charge of the electron] e . . . also appears." After describing some constraints he believes that this new equation must satisfy, he concludes, "I have not yet succeeded in finding a system of equations fulfilling these conditions which would have looked to me suitable for the construction of the elementary electrical quantum and the light quanta. The variety of possibilities does not seem so great, however, for one to shrink from this task."

[3] Ritz was a promising young Swiss physicist whom the Zurich search committee had actually preferred to Einstein, even after Adler withdrew. Ironically Kleiner had described Ritz, not Einstein, as "an exceptional talent, bordering on genius." Ritz was denied the position only because he was terminally ill with tuberculosis, which was to claim his life by July of 1909.

[4] This frank statement elicited a request from Planck that Einstein make it clear that Planck himself was aware of this problem, which Einstein did in a rather awkward addendum to the paper.

Einstein was now focused on finding the "elementary" theory that he expected would underlie both electromagnetic and atomic phenomena, the theory that would put solid walls on the framework of relativity theory, which on its own could go no further in explaining reality at the molecular scale. He communicated his new mindset to Laub in June of 1909: "I am ceaselessly concerned with the constitution of radiation. . . . This quantum question is so important and difficult that everyone should be working on it."

CHAPTER 16

CREATIVE FUSION

"I am very sorry if I have caused you distress by my careless behavior. I answered the congratulatory card your wife sent me on the occasion of my appointment too heartily and thereby reawakened the old affection we had for each other. But this was not done with impure intentions. The behavior of your wife, for whom I have the greatest respect, was totally honorable. It was wrong of my wife—and excusable only on account of extreme jealousy—to behave—without my knowledge—the way she did." In June of 1909, Einstein sent this apology to someone he had never met, George Meyer, who happened to be the husband of Anna Meyer-Schmid, a woman with whom Einstein had flirted on vacation more than a decade earlier. Anna had read in a newspaper of Einstein's appointment as professor in Zurich and had sent him a congratulatory postcard. He had responded with a warmth that certainly could have been misconstrued (or indeed correctly construed): "Your postcard made me immeasurably happy. I . . . cherish the memory of the lovely weeks that I was allowed to spend near you. . . . [I] am sure that you have become as exquisite and cheerful a woman today as you were a lovely young girl in those days. . . . If you are ever in Zurich and have time, look me up . . . it would give me great pleasure."

Up to this point Einstein's relationship with his wife had not entered his correspondence in any problematic fashion, and indeed he described her as a dutiful and supportive wife and mother. This was beginning to change. Einstein was becoming well known, if not yet famous beyond the cognoscenti, and as attention and accolades came his way, he began to perceive Maric as a brooding, negative, and jealous figure. The incident just recounted suggests that there are likely two sides to this story. Einstein's reply to Anna elicited a letter from her that apparently had similar overtones and that somehow came to Maric's attention. She then wrote directly to Anna's husband, rather pathetically claiming that Anna's "inappropriate" letter had outraged Einstein, leading finally to Einstein's apologetic follow-up with its harsh characterization of his own wife's behavior. From Mileva's viewpoint, success had disrupted the harmony of their Bohemian household. "With that kind of fame, he does not have much time left for his wife," she wrote to a friend; and later, "I am very happy for his success, because he really does deserve it. . . . I only hope that fame does not exert a detrimental influence on his human side." In fact Einstein maintained his geniality, modesty, and sense of humor through all his successes, but he did not seem to regard loyalty and attentiveness to a succession of female companions as part of his moral code. Over the next four years Einstein's relationship with Mileva would deteriorate further, leading ultimately to separation and a drawn-out divorce.

Einstein's elevation from the patent office to the status of a university professor, a position of enormous respect in Swiss society, was indeed almost unprecedented. A colleague who was present when Einstein submitted his resignation as technical expert second class recounts that his superior refused to believe that he was leaving to become a professor at Zurich, yelling at him roughly, "That's not true, Herr Einstein. I just don't believe it. It's a very poor joke!" But despite his new prestige, Einstein hardly became a typical Herr Professor of the Weber variety. He lectured quite informally, encouraging his students to interrupt and ask questions, and routinely invited them to further discussions at the Terasse, a café nearby, or even at his home (where he would brew the coffee himself). His students very much appreciated this style, which, despite a certain degree of disorganization, showed his personal concern for them. And by this point he had become quite a good teacher, although he was very surprised at how much time this took from his own research. By November of 1909 he was already confiding in Besso that "my lectures keep me very busy so that my *actual* free time is less than in Bern." This tension between the personal pedagogical relationships, which he enjoyed, and his over-riding drive to confront the mysteries at the frontiers of science ulti-mately would lead him to escape from mandatory teaching when the opportunity presented itself. He never found anything as satisfying as sitting at his desk with a pad of paper serving as window into the universe. And in 1909 nothing was more puzzling and obscure in that window than the "shape" of radiation and quanta.

His January 1909 paper on the current status of the radiation prob-lem, responding as it did to that of Lorentz and others, gave Einstein a natural opening to initiate a scientific correspondence with the man he most admired. At the end of March 1909 he wrote to Lorentz:

> I am sending you a short paper on radiation theory, which is the trifling result of years of reflection. I have not been able to work my way through to a real understanding of the matter. But I am sending you the paper all the same, and even ask you to take a quick look at it, for the following rea-son. The paper contains several arguments from which it seems to me to follow that not only molecular mechanics, but also Maxwell-Lorentz's electrodynamics cannot be brought into agreement with the radiation formula. . . . I cherish the hope that you can find the right way.

A couple of weeks later he followed up with an almost obsequious note praising the "beauty" of Lorentz's derivation of the Jeans law in his Rome lecture and saying that reading it was "a real event." Given Einstein's unshakable insistence that the Jeans law was the only pos-sible outcome of classical reasoning (stated repeatedly since 1905 and reiterated in his most recent paper), one has to wonder if for once he was engaging in a bit of flattery. At any rate he was rewarded a month later with a lengthy reply from Lorentz, with such detailed scientific arguments that Einstein opined "it is a pity" that this "clear and beau-tiful exposition . . . will not be read by all of those who are working in this matter."

Lorentz begins by pointing out that the Planck energy quantization rule, $\varepsilon_{\text{quant}} = h\nu$, makes no sense when it is applied to electrons instead of molecules. Molecules at that point were known to be combinations of atoms bound together, which naturally vibrate when they interact with light; in contrast, electrons in solids are often moving freely as in a gas, and "their existence [free electrons] in metals can hardly be denied." Because of this free, nonperiodic motion, "there can be no question of a definite frequency v, and thus an energy element hv." Thus the quantum of action, h, is not a property of matter in general, but more plausibly one should "[ascribe] to the ether, and not to [matter,] the property that energy can be taken up and given off only in definite quanta."

Note that Lorentz continues to speak of electromagnetic fields as sustained by the ether, the notion Einstein hoped to banish with relativity theory; and he even puts in a mild dig at Einstein and the relativists: "If one regards h as a constant of the ether, one then deprives this medium of part of its simplicity, and directly opposes the views of those physicists who want to deny to ether almost all 'substantiality.'" He then zeros in on the notion of "pointlike" energy packets of light, Einstein's light quanta. Interference of light waves is observed to occur over millions of oscillations, corresponding to light moving a distance of almost a meter—thus he concludes that this is the minimum length of the hypothetical quanta. By a similar argument relating to the focusing of light through a telescope objective, he argues that their width would also have to be macroscopic. "The individuality of each single light quantum would be out of the question," he concludes, wrapping up with the meager consolation, "it is a real pity that the light quantum hypothesis encounters such serious difficulties, [as it] is very pretty, and many of the applications that you and Stark have made of it are very enticing."

Finally, Lorentz throws cold water on Einstein's suggestion that perhaps a modification of Maxwell's equations will explain all these conundrums: "as soon as one makes even the slightest change in Maxwell's equations, one is faced, I believe, with the greatest difficulties." Having rejected all of Einstein's new ideas in the most supportive and gentle tone possible, he finishes with, "permit me to say how glad I am that the problems of radiation theory have given me a chance to enter into a personal relationship with you, after having admired your papers for such a long time."

FIGURE 16.1. Einstein with Hendrik Lorentz circa 1918. Museum Boerhaave Leiden.

Einstein seems to have been so pleased to be noticed by the great man that he was hardly fazed by Lorentz's dismissal of his ideas. Shortly after receiving the reply, he wrote to Laub in starstruck tones: "I am presently carrying on an extremely interesting correspondence with H. A. Lorentz on the radiation problem. I admire this man like no other; I might say, I love him." But this love would not deter him at all from his project, which was to change Maxwell's equations so as to encompass light quanta in some form. In the same letter he continues, "My work on light quanta is proceeding at a slow pace. I believe I am on the right track. . . . But I haven't yet gotten far."

On May 23, 1909, he wrote back to Lorentz an almost equally long and technical letter explaining his program. He is of course "delighted" with Lorentz's detailed letter, which should be read by the whole community; but he then points out that Lorentz's argument that electrons lack a definite frequency of motion is moot. Since we don't know yet the new basic laws of molecular mechanics and electromagnetism, there can be no contradiction with Planck's law, only "the difficulty of generalizing Planck's approach." Einstein has correctly perceived that quantum theory is going to be a general mechanical theory, one implication of which is $\varepsilon_{\text{quant}} = h\upsilon$ for Planck's oscillators, but that some different relation will apply to free electrons. He then goes on to describe the general sort of theory of light quanta that he hopes to develop. In this theory light is neither a particle nor a wave; instead it consists of pointlike objects that carry an extended field along with them, and this field is essentially the conventional electromagnetic field. The pointlike objects are "singularities" in the field, where it becomes infinite, similar to the manner in which the electric field becomes infinite near a single-point electric charge. Every time light is absorbed, one of these pointlike objects disappears and deposits energy $h\upsilon$, where υ is the frequency of the light field. We see here the contrast between the flexible thinking of Einstein and the more rigid views of the aging master, Lorentz, who is wedded to the old categories of physics. While in mathematical detail Einstein's conception is not very close to our modern theory, it is the first, albeit groping, attempt to introduce mathematical objects into physics that are simultaneously particulate and wavelike, a foundation stone in the conceptual structure of modern quantum theory.

Einstein concludes his reply to Lorentz with, "I consider it a great blessing to be able to enter into a closer relationship with you." Here the surviving correspondence with Lorentz from this period ends.

Einstein's exhilaration in the exchange with Lorentz appears to be that of a chess master playing at such a high level that he was delighted to have finally found an opponent who could "give him a game," someone against whom he could test his mettle. But Lorentz could not shake Einstein's conviction that wave and particle properties of light coexist and that a new electrical and mechanical theory would provide a synthesis of these old categories. At this time Einstein was the only scientist on the planet to perceive this need. Of the few who understood both Planck's derivation of the radiation law and Einstein's 1905 work, almost all rejected the notion that the blackbody law implied some form of light quantum. Lorentz and Planck, the two most knowledgeable actors, now agreed that "energy elements . . . play a certain role in the laws of thermal radiation" but also that the notion of physically localized quanta should be rejected.

Johannes Stark, the one established physicist to champion the idea of light quanta at the time, seems to have regarded quanta and wave theory as irreconcilable. Stark, a talented experimenter who would win the Nobel Prize in 1919, was a difficult man who tried to compete in theory beyond his level of competence and became embroiled in disputes with others about the validity of his theories and about scientific priority.[1] However, in April of 1907 it was he who had offered Einstein his first paid academic job, as an assistant in his lab, which Einstein declined for financial reasons (it paid much less than the patent job). And it was Stark who invited Einstein to do the important review article on relativity theory in which the principle of equivalence was first announced. Ironically, this first reputable supporter of Einstein's quantum ideas would become, many years later, a leader, with Philip Lenard, of the Nazi movement against "Jewish physics." All this was in the distant future in July of 1909, when Einstein wrote to his ally, Stark, relating that Planck "stubbornly opposes material (localized) quanta." "You cannot imagine," he continues, "how hard I have tried

---

[1] A year later Einstein gave a blunt analysis of one such squabble in a letter to Laub: "The quarrel between Stark and Sommerfeld is unedifying. Stark has once again produced unadulterated rubbish . . . nothing sensible ever comes out of a quarrel."

to contrive a satisfactory mathematical formulation of the quantum theory. But I have not succeeded thus far." Despite his frustration, he is "[looking] forward to making [Stark's] acquaintance in Salzburg."

In fact, at that time, Albert Einstein, the paradigm-shattering conjurer of modern physics, *still* had not met a single one of the leading physicists of Europe, not Planck, nor Lorentz, nor Sommerfeld, nor Wien (all of whom had corresponded with him). That was about to change. Once a year, since 1822, the physicists of the German-speaking countries assembled in a different city at a major convention of more than 1,300 scientists and physicians, the Deutsche Gesellschaft der Naturforscher und Arzte. In 1908 the assembly had convened in Cologne, and Einstein had planned to attend, but exhaustion prevented him from using his meager vacation time from the patent office for this trip. He had been a star in absentia; it was at that meeting that the mathematician Minkowski had added four-dimensional frills to Einstein's relativity theory and announced the "union" of space and time into a single space-time continuum. But finally, in 1909, with the meeting to be held in Salzburg, Einstein had been invited (probably by Planck) to give a plenary lecture; the new messiah would finally be seen and heard by the congregation and make the acquaintance of the elder prophets. Wolfgang Pauli, who would lead the next generation of quantum theorists, termed this event "one of the turning points in the evolution of theoretical physics."

When Einstein stepped to the podium on the afternoon of September 21 there were over a hundred colleagues in attendance, including Planck, who chaired the session, Wien, Rubens (the blackbody experimenter), and Sommerfeld, as well as younger physicists who would become friendly with Einstein, such as Max von Laue, Max Born, Fritz Reiche, and Paul Epstein. Planck apparently expected him to review the now–accepted relativity theory, but he had chosen a title consonant with his current interests, "On the Development of Our Views concerning the Nature and Constitution of Radiation." He begins his lecture in familiar territory: "Once it has been recognized that light exhibits the phenomenon of interference . . . it seemed hardly doubtful . . . that light is to be conceived as wave motion." This wave motion seems to require an ether in which to propagate, he notes, quoting an authoritative text that designates the existence of the ether as a "near certainty" (we have seen that both Lorentz and Planck are still using the ether concept). But Einstein is having none of it: "However, today we must regard the ether hypothesis as an obsolete standpoint."

Having rudely dismissed the central dogma of electromagnetic theory for forty years, he says it is "undeniable" that certain properties of light suggest a particulate nature. Then, the bombshell:

> It is therefore my opinion that the next stage in the development of theoretical physics will bring us a theory of light that can be understood as a kind of fusion of the wave and emission [particle] theories of light. To give reasons for this opinion and to show that a profound change in our views on the nature and constitution of light is imperative is the purpose of the following remarks.

These remarks are prescient; exactly such a "fusion" theory of light would arise, but it would take almost sixteen years for its earliest forms to emerge, and no other physicist would share Einstein's vision of the future of physics for almost as long.

Having already nailed his revolutionary thesis to the church door, Einstein proceeds with the sermon. He briefly reviews how relativity theory has made the ether superfluous and has suggested that light is an independent entity, not a disturbance in a medium. However, he makes it clear that relativity theory does not itself require light quanta. "Regarding our conception of the structure of light . . . the theory of relativity does not change anything. It is nevertheless my opinion . . . that we are at the threshold of highly significant developments. . . . What I shall say is . . . the result of considerations that have not yet been sufficiently checked by others. If I nonetheless present these considerations, this should not be attributed to excessive confidence in my views but rather to the hope that I may induce one or another among you to concern himself with the problem in question."

He then goes on to review the many processes, such as the photoelectric effect, that seem to depend on the frequency of light and not on its intensity, an elaboration of the same considerations that began in his 1905 paper. This leads him to Planck's derivation of the black-body law and the mysterious "energy element," where he again minces no words: "to accept Planck's theory means plainly to reject the foundations of our [classical] radiation theory." But Planck's theory, he points out, has just been verified yet again by the measurements of Rutherford and Geiger on the value of the elementary charge (the event on which Arrhenius had based his Nobel nomination of Planck a year earlier). Since we cannot reject the Planck law, we must interpret it through quanta of light.

In a Socratic flourish, Einstein then puts on the brakes. "Isn't it conceivable that Planck's formula is correct, but that nevertheless a derivation of it can be given that is not based on an assumption as horrendous-looking . . . ?" His answer to the imagined conservative (who could well have been Planck himself) is a resounding negative. He has a trick up his sleeve, an argument that for the first time does not tend to prove that light is a particle or to prove that light is a wave. He has an argument to prove it is both.

His argument is based on another of his ingenious thought experiments. He imagines a cavity that contains a perfect gas at temperature T and, necessarily, thermal radiation, distributed in frequency according to Planck's law, in equilibrium with that gas (i.e., having the same temperature). In this blackbody cavity is suspended a perfectly reflecting mirror on a rail that can thus move freely in the direction perpendicular to its surface. This mirror is in contact with the gas, and it will suffer collisions with gas molecules at irregular intervals, causing it to move randomly to the left or right.[2] But the mirror also experiences another force, which is due to the pressure of the thermal radiation reflecting off each of its surfaces. This force of radiation pressure is not a quantum effect and had been known since the time of Maxwell; it is responsible for the tail on comets (as hypothesized by none other than

---

[2] This random momentum should correspond to an average kinetic energy of motion $kT/2$ due to the equipartition theorem; the mirror is assumed macroscopic so that the quantum effects, which violate equipartition, can be neglected.

Arrhenius). In the context of Einstein's mirror experiment it has an interesting property: "The forces of pressure exerted on the two sides are equal if the plate is at rest. However, if it is in motion, more radiation will be reflected on the [front surface] than on the back surface. The backward-acting force of pressure . . . is thus larger than the force of pressure acting on the back surface," leading the plate to experience "radiation friction" opposing its motion.

But this cannot be the only effect of the radiation, because if it were, the mirror would be continually taking energy out of the gas through the molecules' collisions with the mirror and effectively transferring it to radiation, hence heating up the radiation, something we know from experience does not occur. The resolution of this paradox is that the radiation, besides it average tendency to slow down the plate, has the same kind of irregular fluctuations in its interaction with the plate as do the gas molecules, transferring back to the plate as much energy as it is receiving on average. It is easiest to imagine this force as arising from the random arrival of unequal numbers of light quanta on each side of the plate, just as the force due to the gas arises from collisions with unequal numbers of molecules on either side. It looks like another argument for light quanta is about to emerge. But this is not where Einstein is heading. Assuming that the Planck law is correct, Einstein can calculate the fluctuating force of radiation, and to the surprise of the crowd[3] he obtains *two* contributions to the force. The first is indeed just what one would expect from "molecules" of light with energy $h\nu$ (and momentum $h\nu/c$, where c is the speed of light) hitting each side of the mirror randomly. But the second term is precisely what Maxwell would have expected, fluctuations in radiation pressure arising from complicated *interference* between light *waves* moving in both directions.[4] There you have it: the magician has removed the screen to show that the lady he has just sawed in half is in fact whole again. Wavelike

[3] Fritz Reiche, a young physicist in the audience, recalls: "I was very much impressed by the second term in the fluctuation formula. . . . I remember of course that people were opposed and tried to find another reason."

[4] Einstein interprets this term as due to Maxwell waves but doesn't derive it from Maxwell's equations; shortly afterward Lorentz himself filled in this step.

and particle-like effects *coexist in the same formula*, a formula that follows simply from the existence of equilibrium between matter and radiation, and the undeniable validity of Planck's radiation law.

Einstein concludes by sketching his "singularity" picture for a new theory of radiation, while admitting "it has not yet been possible to formulate a mathematical theory of radiation that would do justice to the undulatory structure and . . . quantum structure [of radiation]. . . . All I wanted is briefly to indicate . . . that the two structural properties (undulatory and quantum) simultaneously displayed by radiation . . . should not be considered as mutually exclusive."

Planck, as chairman, rose to moderate the discussion of Einstein's paper. One can imagine a hush of expectation: how would the great man respond to this radical proposal concerning the nature of light, from the physicist whose work on relativity Planck had promoted and lauded so widely. A hint emerges as Planck thanks the audience for "listening with greatest interest . . . even where opposition may have emerged." He continues, "Most of what the lecturer has been saying will not meet with any disagreement. I too emphasize the necessity of introducing certain quanta . . . the question is where to look for those quanta. According to . . . Mr. Einstein, it would be necessary to conceive . . . [of] light waves themselves as atomistically constituted, and hence to give up Maxwell's equations. This seems to me a step which in my opinion is not yet necessary. . . . I think that first of all one should attempt to transfer the whole problem of the quantum theory to the area of the interaction between matter and radiation." The voice of conservative reason has been heard; we are not ready for photons, and certainly not for wave-particle duality.

Much later a young physicist who was in the audience, Paul Epstein, was questioned as to the persuasiveness of Einstein's Salzburg lecture. "It had no [great effect]," Epstein replied. "You see the chairman of the meeting was Planck, and he immediately said that it was very interesting but he did not quite agree with it. And the only man who seconded it was Johannes Stark. You see, it was much too far advanced."

CHAPTER 17

THE IMPORTANCE OF BEING NERNST

I visited Prof. Einstein in Zurich. It was for me an extremely stimulating and interesting meeting. I believe that, as regards the development of physics, we can be very happy to have such an original young thinker; a "Boltzmann redivivus [reborn]."

—WALTHER NERNST, MARCH 1910

Although in Salzburg it was clear to observers such as Max Born that "Einstein's achievement received its seal [of approval] before the assembled world of scientists," the achievement most recognized was his theory of relativity, which by then had been well confirmed by the fast-electron experiments of Alfred Bucherer in Bonn. As just noted, Einstein's quantum hypotheses were regarded by all the leaders present in Salzburg as perhaps inventive, but certainly rash and premature. This impression was possibly facilitated by Einstein's decision not to mention in his lecture his lesser-known work on the specific heat of solids. This work, showing that Planck's distribution law of thermal radiation could also be applied to electrically neutral vibrations of atoms, clearly indicated the necessity of a non-Newtonian atomic mechanics, as opposed to treating Planck's "energy element" as some anomaly associated only with the interaction of radiation and matter, as Planck had suggested. But there was one leader, not present at Salzburg, who *was* keenly aware of Einstein's work on specific heat, and to whom it was a very big deal. His name was Walther Nernst.

Nernst was a physical chemist, like Arrhenius, and indeed the two met in their twenties while Arrhenius was studying in Germany in 1886 and became fast friends, as well as boisterous drinking companions. Arrhenius pronounced Nernst's work on heat conduction "the best that any laboratory practitioner has done in a long time" and got him invited to work with his patron, Ostwald, in Leipzig. Like Arrhenius, Nernst was not much to look at. He was short of stature with a "fish-like mouth" and had a distinctive, high-pitched voice that was often employed in claiming priority for some idea or other, which he typically insisted had already appeared in his famous textbook on physical chemistry. When, in the early 1900s, Nernst became well known in Berlin society, a joke circulated about a superman whose brain God had created but whose body had been left to lesser craftsmen. The disappointing result was then brought to life by the Devil for amusement; you can guess the name of this golem.

Nernst himself had a distinctive form of sarcastic humor, which he expressed with a completely deadpan delivery. For example, surrounded by avid hikers including Planck, who climbed mountains well into his seventies, he would opine that he too had climbed a mountain once in his youth, and that would suffice for a lifetime. He was indeed a talented scientist; Einstein praised "his truly amazing scientific instinct combined both with a sovereign knowledge of an enormous volume of factual materials . . . and with a rare mastery of the experimental methods and tricks in which he excelled." Despite what Einstein called his "childlike vanity," he was, according to Nernst trainee Robert Millikan, "in the main, popular in the laboratory, despite the fact that in the academic world he nearly always had a quarrel on with somebody." Millikan recounts that in 1912 he wrote a review chapter for Nernst's famous textbook, dealing with the determination of the electric charge, e (the topic on which Millikan would eventually do Nobel-winning research). Nernst initially accepted the draft, but after Jean Perrin, a well-known French physicist and eventual Nobel laureate himself, annoyed Nernst at a conference, he demanded that Millikan expunge every mention of the man's name from the chapter. Perrin's offense was speaking too long in the lecture before Nernst was scheduled to speak, hence depriving Nernst of some of his allotted time. Such a combination of determination, charisma, and pettiness was Walther Nernst.

These qualities conspired to make Nernst one of the most successful scientific "operators" of the twentieth century, a person capable of convincing officials in government and industry of the importance of scientific work in general, and of his own contributions in particular. Soon these attributes would have a great influence on Einstein's career and on the development of quantum theory. Nernst's political savvy showed itself early in his career. At age twenty-four, in 1888, he developed an idea, originally due to the chemist van 't Hoff, into an important relation now known as the Nernst equation, which allows the useful energy of a chemical reaction to be estimated from electrochemical measurements, and he quickly parlayed this into an academic position at Göttingen. By 1895 he had procured a full professorship and his own physical chemistry institute there, obtained under the threat of leaving for a position in Munich. Around that time he began working on one of the key technological problems of the era: cheap, durable electric lighting. In fact electric lighting provided a primary motivation for the momentous research program on blackbody radiation by Rubens, Kurlbaum, and others that was reaching fruition at almost the same time.

By 1897 Nernst had developed and patented the "Nernst lamp," based on a heated cerium oxide glass rod, which was superior in a number of ways to the incandescent metal filament bulb pioneered by Edison and others. Understanding that the development of the market for lighting was unpredictable, Nernst refused to accept royalties for his invention, which was licensed by the German firm AEG, instead insisting on a large lump sum payment up front. As perhaps Nernst had foreseen, his lamp shone brightly for a few years, magnificently lighting the German pavilion at the Paris Exhibition of 1900 and selling over four million units in the next decade, but eventually losing out to the less-expensive tungsten filament technology perfected by a former Nernst student, the chemist Irving Langmuir. Long before this denouement Nernst had visited Edison in American in 1898, and after being subjected to a lecture on the irrelevance of academic work by the aging inventor, he shouted into Edison's ear trumpet, "How much did you get for your light-bulb patent?" Upon Edison's reply that he got nothing, Nernst bellowed, "I got a million marks[1] for mine! The trouble with you Edison is that you are not a good businessman!"

Because of his negotiating acumen, the commercial failure of his lamp was of little consequence;[2] Nernst was now a wealthy man and a figure to be reckoned with in German society. In 1905 he accepted a professorship at the center of German science, the University of Berlin, where he became a close associate of Planck, whose earlier thermodynamic work had put the Nernst equation on a more sound theoretical footing. His distinguished colleague in Berlin, the organic chemist Emil Fischer, who already had receive the Nobel Prize in 1902, described him approvingly as "versatile, many-faceted, full of curiosity and enterprise." His students referred to him jokingly as the *Kommerzienrat*, German for a successful businessman (in contrast to the usual *Geheimrat*, denoting a distinguished scholar). Even before moving, he had been decorated by the emperor and appointed privy councillor, and upon his arrival he was immediately admitted to the prestigious Prussian Academy of Sciences.

Around this time he made the most important scientific discovery of his life, a principle that is now known as the Third Law of thermodynamics. This principle, in modern terms, is expressed by the statement that the entropy of any system tends to zero as its temperature tends to absolute zero. Unlike the other two laws of thermodynamics, which don't require quantum effects to make sense, this law is all about quantum "freezing." According to Boltzmann, a system's entropy depends on the number of its accessible states at a given temperature.

---

[1] The equivalent of roughly 4.5 million 2008 dollars. Diane Barkan, in her biography of Nernst, states that the actual amount Nernst received is unverifiable, but the legend persists.

[2] One consequence his invention did have, ironically, was the rupture of his friendship with Arrhenius. In 1897 in Stockholm Nernst demonstrated his lamp for Arrhenius, who laughed vigorously when it blew all the fuses in the hotel. From this small incident a lifelong feud nucleated, with the consequence that Nernst received the Nobel Prize only in 1921, after many lesser lights of chemistry had been so recognized.

We already have seen from Einstein's work on vibrations in solids that when a solid gets very cold the atoms cannot vibrate, because they don't have enough thermal energy to reach even the first excited quantum level, which is $h\upsilon$ higher in energy than the lowest (ground) state. In other words the system tends toward a single, unique ground state, and the entropy, by definition, is zero. Nernst's law says this always happens for any system, no matter how complicated.

Why, you may ask, would a chemist with a practical and empirical bent like Nernst care deeply about this apparently abstract principle? Because for some years previously it had been realized that knowledge of a system's low-temperature entropy behavior would allow one to use chemical reaction data over a limited range of temperatures and predict at other temperatures and pressures how much of each product a particular reaction would yield, a fundamental question in all of chemistry. Nernst quickly set out to calculate the "reaction constants" that followed from his principle and showed that they predicted well the results of various experiments. Already in 1910, when he was meeting Einstein for the first time, his approach had been crucially employed by his colleague Fritz Haber to facilitate the development of a chemical technique of immense importance for humanity, the process for removing nitrogen from the air to make ammonia for fertilizer (and explosives).

Characteristically, Nernst referred to the principle he had discovered as "his theorem," which is exactly what it wasn't. A theorem is something logically deduced from other accepted rules or axioms. Nernst's "theorem" was a hypothesis based on analysis of data; it had no grounding in atomic theory at all, nor could it have had at that time. Systems that obey the equipartition principle of classical mechanics do not obey Nernst's law; so his conjecture actually contradicted the currently accepted theory. We now know quantum ideas are essential for the validity of this principle. And that is where Einstein comes in.

Nernst surely realized that his "theorem" was going to require some microscopic underpinning to become a law. And there were absolutely no hints of a microscopic theory that would violate equipartition coming from the leading theorists of the time. Yes, there was that strange business about the radiation law, but it seemed, as Planck would surely have advised him, to relate mainly to matter in interaction with radiation. Thus one can imagine his excitement when he became aware of Einstein's 1907 paper, explaining how applying Planck ideas to vibrations in solids predicted a violation of equipartition, and just the kind of violation (the freezing out of vibrations) that would justify Nernst's "theorem." No wonder that in early March 1910 Nernst made a special trip to Zurich to meet the wunderkind who had provided the first mathematical theory consistent with his historic conjecture.

Despite Einstein's rise in status as an associate professor, very few of his colleagues and acquaintances had any idea that a major intellectual figure was among them. His informality with the students and his ever-present sense of humor hardly signaled an august personage; and his dress had not improved greatly from the time he used a runner from his dresser as an impromptu scarf. His first (and only) doctoral student, Hans Tanner, describes his "rather shabby attire, with trousers too short for him and an iron watch chain," hastening to add that his lecturing style immediately "captured our hearts." An associate professorship for a man with a family could not have supported much elegance anyway; an older colleague at the time wrote of Einstein, to none other than Arrhenius, "I am most interested in the associate professor at the university, A. Einstein, a still young, totally brilliant chap from whom one can learn a lot. . . . I believe he has a great future, at present he lives with his wife and children in very modest conditions. He certainly deserves a better fate."

The general impression of Einstein's status changed suddenly and dramatically after Nernst's visit. George Hevesy, a young assistant at the Zurich Poly at that time, who went on to become a Nobel laureate in chemistry, recalls that Nernst's visit "made Einstein famous. Einstein in 1909 was unknown in Zurich. Then Nernst came and people in Zurich said 'that Einstein must be a clever fellow if the great Nernst comes all the way from Berlin to Zurich to talk to him.'"

Very little is known about the details of this visit except that Nernst filled Einstein in on the state-of-the-art measurements of specific heat as a function of temperature being done in his laboratory, and Einstein surely impressed Nernst with his profound understanding of thermodynamics and statistical mechanics, and with his thoughts on how the quantum hypothesis could clear up many issues. Nernst's reaction is worth quoting at length:

> I visited Prof. Einstein in Zurich. It was for me an extremely stimulating and interesting meeting. I believe that, as regards the development of physics, we can be very happy to have such an original young thinker; a "Boltzmann redivivus [reborn]"; the same certainty and speed of thought; great boldness in theory, which however cannot harm, since the most intimate contact with experiment is preserved. Einstein's "quantum hypothesis" is probably among the most remarkable thought [constructions] ever; if it is correct then it indicates completely new paths both for the so-called "physics of the ether" and for all molecular theories; if it is false, well, then it will remain for all times "a beautiful memory."

The most striking thing about this remarkable quotation is that Nernst, Max Planck's close colleague and friend, refers to *Einstein's* quantum hypothesis without mentioning Planck at all. It is clear to him that in the hands of Einstein, Planck's ad hoc patch-up of radiation theory has become something very different: a vision of a completely new electromagnetic and molecular theory. In fact, there is no indication that Einstein's 1907 proclamation of a sweeping quantum revolution in molecular mechanics was noted by anyone until Nernst took it up, apparently in late 1909 or early 1910. Not a single paper had been written relating to the quantum theory as applied to specific heat between early 1907, when Einstein's work appeared, and February 17, 1910, when Nernst read his first paper on the subject to the Prussian Academy of Sciences, mentioning Einstein's theory briefly at the end.[3] With Nernst leading the charge this changed dramatically; ten such papers appeared in 1911, and over thirty total in the subsequent two

---

[3] "The specific heat decreases strongly at low temperatures . . . corresponding to the requirements of Einstein's theory [that] it tends to zero" (Nernst, February 17, 1910).

years. Moreover it is very likely that, during or shortly after that visit, Nernst conceived the project of bringing Einstein to Berlin; a postcard from Nernst dated July 31, 1910, begins, "I have made inquiries regarding Einstein, but have not yet received any news."

As for Einstein, who by now had been struggling fruitlessly for more than two years to explain light quanta by modifying Maxwell's equations, the visit by Nernst was a great morale boost. A week after Nernst left he wrote to Laub: "For me the theory of quanta is a settled matter. My predictions regarding the specific heats are apparently being brilliantly confirmed. Nernst, who has just been here to see me, and Rubens are busily engaged in the experimental verification, so that we will soon know where we stand." And three months later he wrote to Sommerfeld with further results: "It seems incontrovertible that energy of a periodical nature, wherever it occurs, always occurs in energy quanta that are multiples of $h\nu$ . . . [whether] as radiation or as oscillation of material [molecular] structures. . . . It now seems pretty certain that as regards the heat content, the molecules of solid substances behave essentially similar to Planck's resonators. Nernst found the relationship confirmed in the case of silver and some other substances."

But all this support for his heuristic ideas did not for one minute distract him from the underlying challenge: how do wave and particle properties manage to coexist in a full mathematical theory of quantum mechanics or electrodynamics? With characteristic wit he summarized his view in the same letter to Sommerfeld. "The crucial point in the whole question seems to me to be: 'Can energy quanta and Huygens' principle [of wave interference] be made compatible with each other?' The appearances are against it but, as it seems, the Lord knew all the same how to get out of the tight spot."

CHAPTER 18

LAMENTING THE RUINS

As for knowing, nobody knows anything. The whole story would be a delight to diabolical Jesuit fathers.

—ALBERT EINSTEIN, NOVEMBER 1911

A distinguished white-haired man, impeccably dressed, just over seventy, took the podium to explain his theory. In the small audience of twenty-four, listening attentively, were Einstein, Lorentz, Planck, Nernst, Wien, Rutherford, Madame Curie, Jean Perrin, and H. K. Onnes, all of whom were current or future Nobel laureates, as well as major scientific figures such as Sommerfeld, Rubens, Poincaré, and Jeans. The venue was a small meeting room in the elegant Hotel Metropole in Brussels.

I decided to take as my starting point the one general concept that could meet the demands of the most scrupulous, philosophical and constructive mind: positive and negative ether, atomically and invariably cubifiable. The interfaces between them form alternating positive and negative atomic planes; there is a universal competition between these two different ethers, although they are essentially the same, because of spacifiable and superficialisable molecules. Spacification and superficialisation are energetically produced, and energy is produced exclusively by molecular contacts. Molecular contact, which has hitherto been neglected, is an essential element in my theory. . . . I undertook to create the active Universe with the intimate and well-defined mechanism of its own primitive elements.

Upon his conclusion, none of the assembled dignitaries rose to take issue with either the correctness or the coherence of this eccentric "theory of everything." On the contrary, Lorentz, the admired chairman of the proceedings, took pains to thank the speaker for "the report he was good enough to send us and . . . the talk he has just given us explaining its main ideas." And why had the brilliant men and woman present suspended their otherwise highly developed critical faculties? They were engaging in a time-honored ritual of both art and science: humoring the wealthy patron.

The speaker was the prosperous Belgian industrialist Ernest Solvay, inventor of an efficient and lucrative process for making soda, now an exponent of scientific progress across all spheres of human activity and a dilettante physical theorist. The occasion of his exposition was the legendary First Solvay Congress of 1911, the beginning of the modern specialized science conference. Monsieur Solvay, enlisted through the political acumen of Nernst, had footed the entire bill for this opulent and exclusive gathering of the crème de la crème of European physics. Fortunately, Solvay's own "gravito-materialist" theory did not occupy much of the subsequent discussions of atomic theory, which his scientific assistant said he regarded as too "highly specialized."

Einstein had arrived in Brussels from a very different place—intellectually, professionally, and geographically—than he had been only nineteen months earlier when he had been so delighted and flattered by Nernst's visit to Zurich. From the practical point of view, Nernst's visit marked the beginning of a meteoric rise in Einstein's stature and conditions, culminating in the invitation to deliver the concluding "report" at the Brussels conference, perhaps the most elite gathering of scientists in history. Einstein's associate professor position at the University of Zurich was hardly a sinecure; it barely offered him a middle-class living and was not part of any track that would lead him to the position of an *Ordinarius*, or full professor, who could call his own shots. However, a bare five weeks after Nernst concluded his visit in early March of 1910, Einstein was nominated for a full professor's position at the German University in Prague, a post he was eager to take, despite his comfort with life in Zurich, because it was a necessary step up the ladder of scientific advancement. During the summer of 1910 the appointment stalled briefly as, for the first time, explicit anti-Semitism seemed to block Einstein's career; the cognizant ministry in Vienna decided to appoint the second choice of the search committee, an Austrian, Gustav Jaumann, whose role in the history of science is limited to this one cameo appearance. Fortunately for Einstein, in an expression of wounded pride, memorable for its inaccuracy, Professor Jaumann effectively withdrew his candidacy, accusing the university of "chasing after modernity while being blind to real merit." As a result, by April of 1911, Einstein and family had moved to Prague, where he took up his new position as full professor of theoretical physics.

FIGURE 18.1. Einstein, the newly minted full professor, in Prague, 1912. ETH-Bibliothek Zurich, Image Archive.

This was his status when he appeared at the Solvay Congress in October of 1911, but his circumstances were still fluid, with a steady current upward. Even before he left Zurich, he had been contacted by Emil Fischer, the German Nobel laureate in chemistry, with the extraordinary news that an anonymous industrial donor, likely prompted by Nernst, had offered Einstein a gift of fifteen thousand marks "to promote your [scientific] work," with no strings attached.[1] By now Planck had publicly described the theory of relativity as "in boldness [surpassing] anything so far achieved in speculative natural science . . . [it] can only be compared to . . . [the revolution] produced by the introduction of the Copernican world system." It was unlikely that Prague could

---

[1] The letter from Fischer mentioned only Einstein's "great theoretical papers in the area of thermodynamics" and not relativity theory, indicating the input of Nernst.

keep this rising star for long. Only a few months after Einstein's move there, his close friend Heinrich Zangger, dean of the medical faculty at the University of Zurich and someone whom Einstein admired greatly, began campaigning to bring him back to Zurich by creating a full professor's position at his alma mater, the Zurich Poly.[2] While Einstein was outwardly courteous regarding conditions in Prague, he felt isolated and estranged from the culture, privately confiding that "the air is full of soot, the water life-threatening, the people superficial"; in hindsight he termed the city "semi-barbaric." Thus when Zangger's campaign was successful, Einstein immediately accepted the position. He ended up returning to Zurich, with this much-improved status, in July of 1912, barely a year after moving to Prague. His return was made easier by an act of God: the death of his old nemesis, Weber, which he unsentimentally pronounced "a good thing for the Polytechnic." In the middle of all this professional back-and-forth, Mileva had given birth to his second son, Eduard, in the summer of 1910, notwithstanding which his relationship with his wife continued to deteriorate.

Throughout all this change, the constant in Einstein's life was his scientific work, and in particular his focus on what he considered the most important problem of the day, quantum theory. Nernst's visit to see Einstein in Zurich, and Nernst's high-profile focus on quantum problems, had finally begun to make the physics community aware of what Einstein had known since 1907: there were really two puzzles of quantum theory. The radiation formula, photoelectric effect, and other similar phenomena suggested that light, conventionally conceived as a wave, came in quantized energy units, $h\nu$, and had particulate properties. At the same time the specific heat theory, now corroborated in its essentials by Nernst's experiments, pointed strongly to the proposition that molecular mechanics violated Newton's laws and also involved quantization of energy in units of $h\nu$, at least when the molecular motion was periodic. Since these vibrations could be electrically neutral and thus not interact with light directly, this behavior appeared to be

---

[2] The Zurich Polytechnic had just been raised to the status of a technical university capable of granting doctorates and renamed ETH Zurich.

essentially independent of the quantum properties of radiation. Einstein had a major decision to make: which quantum problem would he focus on? He chose the quantum theory of radiation.

Already in early 1909 Einstein had confided to his idol, Lorentz, that his program was to tinker with Maxwell's equations in order to generate a theory of light quanta. Lorentz, who undoubtedly had the deepest understanding of electromagnetic theory of his generation, had warned Einstein, "as soon as one makes even the slightest change in Maxwell's equations, one is faced . . . with the greatest difficulties." As he toiled onward through 1909 and 1910, Einstein began to realize that Lorentz was prescient; changing Maxwell's equations was like touching up the *Mona Lisa*. Not frivolously had Boltzmann proclaimed of these equations, "was it God that wrote those lines?" The mathematical structure was perfect, and every modification implied a contradiction with the multitude of known electromagnetic phenomena. Einstein was in the midst of these difficult explorations in September of 1909 when he admitted in the conclusion of his Salzburg lecture, "it has not been possible to formulate a mathematical theory of radiation which will do justice both to the undulatory structure and the . . . quantum structure. . . . [The] fluctuation properties of radiation [that he had demonstrated in his lecture] . . . offer few formal clues on which to build a theory." He boldly prophesied the emergence of a fusion theory of radiation, yet had nothing to show in support but his vague notion of "singularities" attached to an extended field, which he quickly qualified with the statement, "no importance should be attached to such a picture as long as it has not led to an exact theory."

In his correspondence throughout the year after Salzburg lecture, Einstein repeatedly alludes to his struggles with radiation theory. On New Year's Eve of 1909 he writes again to Laub, who was his constant sounding board during this period: "I have not yet arrived at solution to the light quanta question, but I have found quite a few significant things while working at it. I'll see whether I might not yet succeed in hatching this favorite egg of mine." In January of 1910 in a letter to Sommerfeld he describes his increasingly radical hypotheses to explain the dual properties of radiation: "maybe the electron is not to be conceived as such a simple structure as we think? There is nothing one would not consider when one is in a predicament." By July of that year, buoyed by Nernst's visit, he writes again to Sommerfeld, confident that the two quantum hypotheses (for light and for matter) are correct but allowing that no progress has been made reconciling quanta and waves. He has decided that "[a] crudely materialistic conception of the point structure of radiation . . . cannot be worked out." On August 2, to Laub, shortly after being visited by Sommerfeld, he writes: "I have not made any progress regarding the question of the constitution of light. There is something very fundamental at the bottom of it." Then, again to Laub, in November of 1910, a ray (or wave?) of light: "At the moment I am very hopeful that I will solve the radiation problem, and that I will do so without light quanta. I am awfully curious how the thing will turn out. One would have to give up the energy principle [conservation of energy] in its current form." Alas, one week later: "The solution of the radiation problem has again come to naught. The devil played a dirty trick on me."

Unfortunately, none of the notes containing Einstein's failed attempts to quantize radiation have survived, and we know very little about the mathematical structures he played with and discarded during this period. This letter is, however, the last time that Einstein suggests he knows how or is close to solving the quantum puzzle for radiation. In December of 1910, again he expresses frustration: "the riddle of radiation will not yield . . . the secret remains unsolved."

In hindsight we know that Einstein was on the wrong track in trying to change Maxwell's equation to encompass quanta. Einstein was trying to do the most natural thing: find a new set of electromagnetic equations that contained the fundamental quantum constant, h, either explicitly or perhaps, he thought, implicitly, through the ratio e^2/c (the electron charge squared divided by the speed of light), which has the same units as Planck's constant. All the other known constants of nature at the time (except h and e) appeared in the known laws of nature. The speed of light appears directly in Maxwell's equations; the gravitational constant, G, in the Newtonian law of gravity; Boltzmann's constant, k, in the entropy principle, $S = k \log W$. As these constants appear in the fundamental laws, one can calculate from these laws consequences that depend on the values of c or G or k.

Planck's constant was not introduced in connection with a new physical law; its introduction was an ad hoc insertion in the midst of evaluating the blackbody entropy from Boltzmann's principle. Einstein's hypothesis of light quanta had the same ad hoc character, essentially inherited from Planck. That is why Einstein repeatedly emphasized that "the so-called quantum theory of today is, indeed, a helpful tool but . . . it is not a theory in the usual sense of the word." He naturally assumed that a quantum theory of radiation would require a generalization of Maxwell's equations, which would introduce Planck's constant but would reduce to the original equations in contexts where no hint of quantum behavior was observed.

But the deck was stacked against him, and his assumption was wrong for a most subtle reason. Maxwell's equations remain true and unchanged in quantum theory; it is only their *interpretation* that changes: they are the *wave* equation governing the dynamics of the photon, in the same sense that eventually the Austrian physicist Erwin Schrödinger would discover the quantum wave equation describing the dynamics of the electron. But there is a decisive difference between the two equations: right in the middle of Schrödinger's equation for the electron, planted like a flag, is Planck's constant, just as Einstein had expected in an analogous equation for photons. Einstein had picked the short straw. In the radiation problem the "quantum of action," h, was hidden in plain sight, invisible because of the principle of relativity, Einstein's own creation!

If there were indeed particles of light, they could have no inertial mass, as no massive object can reach the full speed of light. Nonetheless, as we have seen, radiation can exert pressure (i.e., it can transfer momentum) even though it is massless.[3] The relativistic relation between the energy, E, of a light wave and its momentum, p, is $E = pc$; this relationship is embedded in Maxwell's differential equations. In

[3] In Newtonian physics the momentum of an object is its mass times its velocity ($p = mv$), and would be zero for a massless object.

quantum theory, as we now understand it, photon energy is quantized and proportional to h, *but so is momentum*; so the factor h cancels in the equation relating the two. This is the essential reason that h does not appear in Maxwell's equations. It "should" be there, but it cancels because photons are massless. This does not happen for massive particles, which thus are governed by quantum equations in which h sticks out like a sore thumb. We now believe that photons are the *only* freely propagating massless particles in our universe, so Maxwell's equations are the only quantum equations where h does not appear explicitly. How's that for bad luck? If Einstein had instead decided to focus on the mechanics of electrons in atoms, as would Niels Bohr in a just a couple of years, perhaps the history of physics would have been different. But his quixotic search for the quantum version of Maxwell's equations defeated him, and he soon would lay down his lance.

By May of 1911, writing from Prague, a new note of resignation appears in a letter to Besso. "I no longer ask whether these quanta really exist. Nor do I try to construct them any longer, for I now know that my brain cannot get through in this way. But I rummage through the consequences as carefully as possible so as to learn the range of applicability of this conception." And then, just as Einstein is abandoning his four-year struggle, comes the invitation:

Dear Sir,

To all appearances, we are at the moment in the midst of new developments regarding the principles on which the classical molecular and kinetic theory of matter has been based. . . . Messrs. Planck and Einstein have demonstrated that . . . contradictions disappear if one imposes certain limitations on the movement of electrons and atoms, . . . but this interpretation in turn . . . would necessarily and indisputably entail a vast reform of our current fundamental theories.

To that end, the undersigned proposes to you to participate in a "Scientific Congress" which will . . . bring together in a small meeting, several eminent scientists. . . . I hope that I can count on your collaboration, and I beg to assure you, dear, Sir, of my highest esteem.

Signed: Ernest Solvay, June 9, 1911

One can just imagine Einstein's reaction as he read the letter: Tell me about it, buddy. The invitation was of course too prestigious to turn down, and he feigns great interest in his reply to Nernst a few days later. He has been asked to give the report on the current status of the problem of specific heat, and he accepts. But his heart, and his fertile scientific imagination, are elsewhere. By August 1911 his correspondence with Laub contains the first hints of a new passion: "The relativistic treatment of gravitation is causing serious difficulties. I consider it probable that . . . the principle of the constancy of the velocity of light holds only for spaces of constant gravitational potential." Within two weeks of that letter he is corresponding with the astronomer Willem Julius about the redshift of the wavelength of light rays from the sun due to its gravitational[4] field. In September, with the Solvay Congress looming in less than two months, he replies irritably to a letter from Besso: "If my answer is not . . . thorough, it's because my drivel for the Brussels congress weighs down on me."

Thus, while the First Solvay Congress was a point of departure for the field, the quantum problems it addressed would henceforth be the central problems of physics, it marked a temporary surrender for its youngest participant. Einstein's report was thorough and scholarly and explained how sharply contradictory the various modes of reasoning were, as revealed, for example, by the fluctuations in the energy of radiation: "We stand here before an unsolved puzzle, just as in the study of thermal motion in a solid. . . . Who would have the audacity to give a categorical answer to these questions? I only intended to show here how fundamental and deep-rooted the difficulties are in which the radiation formula enmeshes us." He presented no new ideas for how to get out of these difficulties; the optimism of Salzburg had dissipated.

On a personal level Einstein enjoyed the conference greatly; he described how he was "enchanted" with the French trio of Jean Perrin, Paul Langevin, and Madame Curie. Lorentz, whom he had already

---

[4] The correspondence was actually initiated by a job feeler from Julius's University of Utrecht, which Einstein ultimately received and turned down in favor of ETH Zurich.

met for the first time earlier that year, again awed him: "H. A. Lorentz chaired the conference with incomparable tact and unbelievable virtuosity." Planck's integrity won him over: "he is a completely honest man who shows no consideration for himself." But as for science, everyone there was just discovering the forbidding territory he had been surveying for years. His final verdict was delivered to his old friend Besso, for whom diplomacy was not required: "In Brussels, too, they acknowledged the failure of the theory . . . but without finding a remedy. In general the Congress . . . resembled the lamentations on the ruins of Jerusalem. Nothing positive has come out it. . . . I did not find it very stimulating because I heard nothing that I had not known before." In the aftermath he seems to have suspected that his focus on radiation was the wrong path, writing to Lorentz a month later, "the h-disease[5] looks ever more hopeless. *Still I believe that the purely mechanical side will be the first to be cleared up.*" In this he would be proved right.

There was one claim enunciated at Brussels that Einstein alone of all present might have found particularly interesting and worthy of dispute, although he would never have done so in public, because that claim was made by Solvay himself. In the midst of his incomprehensible lecture on "positive and negative ether" Solvay had included the statement, "I took as my starting point Newton's wonderful law [of gravity], which is uncontested and therefore able to satisfy the most rigorous scientific mind." In fact it was just this law that Einstein had now begun to question seriously. By March of 1912 he wrote to a friend, "I'm working at full speed on a problem (gravitation). You should forgive me my long silence." In October of 1912, having returned to Zurich, he declined an invitation from Sommerfeld to speak on quantum theory with the words, "I assure you that I have nothing new to add to the question of quanta that might be of any interest. . . . I am now working exclusively on the gravitation problem." Sommerfeld, who had contacted Einstein with the speaking invitation on behalf of the famous mathematician David Hilbert, wrote in despair to

[5] Einsteinian synecdoche: Planck's constant, h, stands for the entire quantum conundrum.

Hilbert, "My letter to Einstein proved useless. . . . Apparently he is so deeply involved in the problem of gravitation that he turns a deaf ear to all else."

Einstein, at least temporarily, had put up a metaphorical sign over the door to the atom: THIS WAY LIES MADNESS. He admitted as much with a joke he told to a friend and colleague, Philipp Frank, whom he had met during his year in Prague. Frank had quickly been captivated by Einstein's wit and whimsy: "his sense of humor was readily apparent . . . when someone said something funny . . . the laughter that welled up from the very depth of his being was one of his characteristics which immediately attracted one's attention." As Frank tells the story, "about that time Einstein began to be much troubled by the paradoxes arising from the dual nature of light[6] [wave and particle]. . . . his state of mind over this problem can be described by this [the following] incident." Einstein's office in Prague looked over a park, whose patrons had odd characteristics: only women appeared in the morning, only men in the afternoon, sometimes alone and talking to themselves, other times in groups engaged in vehement discussions. The explanation for these patterns turned out to be that the park belonged to the mental asylum of Bohemia, whose less violent patients were allowed its use. Einstein led Frank to his window, and then remarked playfully, "there are the madmen who do not occupy themselves with the quantum theory."

---

[6] Frank was of course wrong here; Einstein had been concerned about this paradox for at least four years prior to moving to Prague.

CHAPTER 19

A COSMIC INTERLUDE

Scientific endeavors are quite extraordinary; often nothing is more important than seeing where it is not advisable to expend time and effort. . . . an instinct must be developed for what is just barely attainable upon the exertion of the utmost effort. [My recent] magnetic experiment, for example, could have been done by any old lout. But general relativity is of another kind. Having actually arrived at this goal gives me the greatest satisfaction of my life, even though up to now not a single colleague in the field has recognized the depth and necessity of this path.
—ALBERT EINSTEIN, MAY 31, 1915

Having decided that a true quantum theory was not yet attainable in 1911, even with his utmost exertion, Einstein devoted himself for the next four years primarily to his new theory of gravity, which arose as a natural generalization of his special theory of relativity and is hence termed the general relativity theory. The inspiration for his first work in this area, the special theory, had come from the properties of electromagnetic waves, arising from Maxwell's equations. Gravity played no role at all in his thinking, and in fact prior to 1907, in contrast to his well-documented interest in atoms, there is no evidence Einstein was particularly interested in gravitational phenomena. Gravitational effects were of primary interest in astronomy, as it is only on the celestial scale that gravitational forces dominate over electromagnetic and nuclear forces. However, Einstein was nothing if not dogged in pursuit of conceptual clarity. Upon enunciation of the principle of relativity, the statement that the laws of physics should appear the same to all observers in uniform relative motion, he quickly noted that "the question arises whether this statement should not also be extended to non-uniform [i.e., accelerated] motion." When in accelerated motion, you feel forces not present in uniform motion. For example, when you go from zero to sixty in four seconds, you are pressed back in your seat; you cannot say you are standing still and the world around you is accelerating backward, because people outside your car are not pushed forward. The conclusion that you are "really accelerating" and they are not seems unavoidable. Thus it appears that one can at least define absolute acceleration, if not absolute velocity.

However, in 1907, Einstein got the idea that pointed the way to a theory in which *all* motion, even that involving acceleration, is relative—a general theory of relativity. The hint to his idea is contained in a phrase used by the zero-to-sixty crowd: "pulling gs." The force you feel pushing you back in your seat feels sort of like gravity; in fact, if your acceleration were constant and exactly the same as that due to gravity at the earth's surface (9.8 meters per second squared, denoted by the letter g), you would "weigh" exactly the same with respect to a scale placed in the back of your seat as you do with a scale standing upright. Einstein surmised that the inertial force experienced during acceleration is completely indistinguishable from the force of gravity. It followed that a general theory of relativity, with no special, privileged states of motion, could perhaps be achieved if it were also combined with a new theory of gravitation. He sketched the earliest, most primitive outline of such a theory in 1907 but did nothing more on it until 1911, as he began to relinquish his single-minded focus on quantum theory and the nature of radiation. For the next four years almost all his original research papers were on the theory of gravitation, and his few works pertaining to atoms and/or quantum phenomena were not of a groundbreaking nature. His final success did not come until November of 1915, six months after his premature declaration of success (quoted above).

This episode in his scientific life was a departure from his earlier modus operandi. Contrary to some depictions, Einstein had been very interested in and aware of experiments in all his previous work; in fact, during this very same year, 1915, he had published his only experimental paper, an attempt to determine the origin of atomic magnetism in collaboration with Lorentz's son-in-law, Wander J. de Haas.[1] Moreover, ever since his student days he had not been much excited about higher mathematics, which he dismissed as needless erudition. He wanted to explain nature, not impress his fellow physicists with his mathematical prowess. General relativity changed his outlook. He quickly realized that he would need more sophisticated mathematical tools than previously, and his theory building was not motivated by any puzzling experiments or even community consensus about fundamental questions, but rather by his own conviction that a consistent framework for physical laws must exist in which all motion is relative. The fact that at the end of a herculean struggle he ended up with a beautiful mathematical construction, which also predicted and explained a few observable astronomical phenomena, created a paradigm of supreme theoretical insight for all who followed.

Nothing quite like this had transpired before in natural science. As early as 1919 J. J. Thomson, the British Nobel laureate. pronounced it "one of the highest achievements of human thought." Quantum pioneers and Nobel laureates Paul Dirac and Max Born went one step further: Dirac called general relativity "probably the greatest scientific discovery ever made," and Born pronounced it "the greatest feat of human thinking about nature." Einstein himself was transported by his breakthrough: "The theory is of incomparable beauty," he told his friend Zangger; to Besso he wrote, "my boldest dreams have now come true." The die was cast; henceforth this achievement would overshadow everything he had done or would do in theoretical physics, not just to the public, who went crazy over curved space and bent light rays when news of the theory's confirmation emerged in 1919, but eventually to Einstein himself. His autobiographical notes, written at his seventieth birthday, present a view of his scientific career in which all his earlier work was the prelude to general relativity and all his subsequent work flowed from it. His quantum mania is barely mentioned.

---

[1] "Experimental Proof of the Existence of Ampere's Molecular Currents," by A. Einstein and Wander J. de Haas, *Koninklijke Akademie van Wettenschappen te Amsterdam*, Section of Science Proceedings, vol. 18, pp. 696–711 (1915–1916).

But this is the perspective of a man who knows how the story will turn out: he knows it will end with an atomic theory he cannot fully accept. The Einstein of 1915 was still in hot pursuit of the new "fusion theory," as he himself had dubbed it, in which light would be both a particle and a wave, and Newtonian mechanics would be replaced by a theory of particle motion that naturally incorporated quantum discontinuity. In February of 1916 he wrote to Sommerfeld, waxing poetic about Sommerfeld's recent theory of atomic spectral lines, which he said had "enchanted me. A revelation!" even calling it, a few months later, "among my finest experiences in physics." By the end of May 1916, six months after completing general relativity, he already was presenting new work on an application of quantum theory to physical chemistry, a proof that any chemical reaction that requires the input of light to proceed, absorbs energy in the amount of $h\nu$ per molecular reaction. By early July of 1916 his next great quantum breakthrough had materialized. He published an initial report that month and was polishing it for a more definitive presentation on August 11, 1916, when he wrote to Besso: "A brilliant idea has dawned on me about radiation absorption and emission; it will interest you." That idea would be the foundation of the modern quantum theory of radiation, the first chapter in every modern textbook on the subject.

The Einstein who wrote those lines was living a very different life from the expatriate professor in Prague who in 1911 had temporarily left quantum theory to the lunatics and raised his intellectual gaze from the atom to the cosmos. Recall that in July of 1912 the newly eminent physicist A. Einstein, father of two young boys and husband of one increasingly unhappy spouse, had returned to Zurich as a full professor of theoretical physics at the Poly, now upgraded in status to the Swiss Federal Institute of Technology (ETH). The Einstein family loved Zurich and had many close friends there, but a variety of forces were conspiring to attract Professor Einstein to a different equilibrium state.

Already, two years before this, the formidable Nernst had begun pulling the levers that would pry Einstein out of Switzerland and into Germany. In this effort he had gained by 1912 a powerful ally in his colleague Fritz Haber, the assimilated German-Jewish chemist who was responsible for the development of the ammonia extraction process, a process that would be critical to Germany's military strength and, ultimately, to the agricultural revolution of the twentieth century. In April of 1912, even before returning to take up the post at Zurich, Einstein had visited Berlin and spoken with Nernst, Planck, Haber, Rubens, and Warburg; and job possibilities were already being floated. As an intellectual center of the new quantum physics, Berlin outshone not only Zurich but any place else on the planet as well, and this clearly interested Einstein. However, with his famous intellectual independence, he hardly needed to move yet again simply to improve his scientific neighbors. But then two new factors entered the equation.

On that very first trip to Berlin in April of 1912 he renewed his acquaintance with a female cousin, Elsa Einstein, whom he remembered fondly from childhood visits. Elsa, a divorcée, three years his senior with two daughters from her marriage, was Albert's cousin through both her father's and mother's family; but, by the customs of the time, this would not have hindered at all an amorous relationship between the two. Later Elsa would claim that she had fallen in love with Albert when she knew him as a boy because of his beautiful violin playing. She was as comfortable and familiar to Einstein as Mileva had been different and intriguing. Born in the Einstein ancestral home of Hechingen,[2] she shared his Swabian dialect and prized the Swabian value of gemütlichkeit, which narrowly refers to a cozy domestic environment but more generally denotes warmth, good humor, and acceptance. While contemporaries judged her prettier than Mileva, she was not a beauty, but shared Einstein's stocky build and curly hair to the extent that later in life people commented on how much they looked alike. Einstein was drawn to her, it seems, as someone who would understand his needs and take care of him, which in fact is exactly what transpired.

How Elsa and Albert reconnected on that first visit to Berlin in April 1912 is not known in detail, but their reunion had an immediate effect on both. Within days of his return Elsa wrote to Einstein at his work address (to avoid Mileva's surveillance). While Elsa's letters from

[2] Einstein's "certificate of presidency" for the Olympia Academy (figure 8.1) had labeled him "the man of Hechingen."

the time have not survived, Albert's were lovingly preserved by Elsa, who clearly saw the prospect of a rewarding future life with him. In his reply to her first letter he wrote, "I can't even begin to tell you how fond I have become of you during these few days. . . . I am in seventh heaven when I think of our trip to Wannsee [the forest near Berlin]." In this very first letter he seems already to have fixed his mind on a future romantic relationship, continuing quite boldly, "I have to have someone to love, otherwise life is miserable. And this someone is you; you cannot do anything about it, as I am not asking you for permission." And, finally, he warns her not to assume that he is unmanly because he defers to Mileva in public: "Let me categorically assure you that I consider myself a full-fledged male. Perhaps I will sometime have the opportunity to prove it to you."

During the same visit to Berlin at which he kindled the relationship with Elsa, Einstein was courted for a position in the Physikalisch-Technische Reichsanstalt, the laboratory at which many of the fundamental studies on blackbody radiation had been done. However, this position did not materialize and in any case would not have held much attraction for Einstein. In fact he had told Elsa in that first letter that "the chances of my getting a call to Berlin are, unfortunately, rather slight." And, perhaps for this reason, Einstein rapidly backtracked on his overture to Elsa, saying just three weeks later, "it will not be good for the two of us as well as for others if we form a closer attachment." However, his judgment on his prospects in Berlin was overly pessimistic; by July of the next year (1913) Walther Nernst and Max Planck had traveled to Zurich to make him an offer he could not refuse.

Unless they were independently wealthy, as in the case of Maxwell and Lord Rayleigh, even the greatest physicists had to earn their keep by teaching, administration, or supervision of projects of practical utility. However, Nernst, Haber, and Planck had convinced the Prussian authorities that Einstein would be such a jewel in their crown that he should be an exception. He was offered a professorship at the University of Berlin with no teaching duties (except by his own choice) as well as the directorship of a new theoretical physics institute, which did not yet exist and which he could run with as little

FIGURE 19.1. Einstein playing the violin along with the daughter of his ETH colleague Adolph Hurwitz, who is conducting, in 1913. Einstein's triumphant return to Zurich would be short-lived, as he was soon lured away to Berlin. ETH-Bibliothek Zurich, Image Archive.

structure as he saw fit. And, finally, he was to be elected to the august Royal Prussian Academy of Sciences as its youngest member.[3] This last honor was considered so great that when he told the news to an esteemed older colleague at Zurich, Aurel Stodola, it brought a "tear of joy" to Stodola's eyes, "because ideal justice had been meted out to somebody on this earth."

It is not likely that Einstein cared for this honor nearly as much as others did. Philipp Frank recounts a jest Einstein made when a colleague at Zurich remarked that admission to the academy always came so late

---

[3] The nominators, Planck, Nernst, Rubens. and Warburg wrote: "The undersigned are aware that their proposal to accept so young a scholar as a regular member of the Academy is unusual; but they are of the opinion not only that the unusual circumstances adequately justify the proposal, but also that the interests of the Academy really require that the opportunity . . . to obtain such an extraordinary person be taken advantage of." The academy also took the highly unusual step of granting Einstein a generous salary for this nominally honorific post.

in life that it no longer made the honoree happy; Einstein replied that then he should be eligible for immediate admission "because it would not make me happy even now." But while Einstein was unimpressed by such vanities, the offer of complete intellectual freedom came at the perfect time. He was halfway through his historic struggle with the theory of gravitation, aware that he was on the verge of changing mankind's view of the universe, but not at all confident of success. Moreover he was closely following the developments in quantum theory, where Berlin was the center. At the same time, he found his teaching duties at ETH increasingly taxing, draining time from his epoch-making research. And now the offer came to be simply left alone in Berlin to think as he pleased! To the consternation of Mileva, he accepted the offer.

In speaking to scientific colleagues, Einstein always cited freedom to focus on his research as his motivation for moving. To Lorentz he wrote, "I couldn't resist the temptation of a post in which I would be free from all obligations and be able to indulge wholly in my musings." To Ehrenfest he spoke more plainly, "I accepted this odd sinecure because lecturing in class was getting oddly on my nerves, and there I won't have to do any lecturing." All the perquisites for this life of untroubled scientific contemplation were put into place as promised by Nernst and Planck, and by April of 1914 Einstein had arrived in Berlin, traveling separately from his family. Barely two months later Mileva and his sons had come and gone, returned to Zurich. Mileva had "howled unceasingly" against the move and hated the city at first sight. Their relationship had become frigid and hostile, and her departure began a separation from him, which would prove permanent, although not accepted as such for many more years. Einstein wept copiously at the train station seeing them off and had to be supported by Haber as he walked home. For years he bemoaned the estrangement from his children but nonetheless quickly adapted to a new, less-encumbered lifestyle in Berlin. Barely a year later he wrote to his old friend Zangger, "In personal respects I have never been so at peace and happy as now. I am living a very secluded and yet not lonely life, thanks to the loving care of my cousin, *who had drawn me to Berlin in the first place, of course*" (italics added).

CHAPTER 20

BOHR'S ATOMIC SONATA

Europe in its madness has now embarked on something incredibly
preposterous. At such times one sees to what deplorable breed of brutes we
belong. I am musing serenely along in my peaceful meditations and feel only
a mixture of pity and disgust.
                    —ALBERT EINSTEIN, AUGUST 19, 1914

Einstein had not only changed his domestic situa-
tion by moving to Berlin; he had radically changed his social and political environment. Within a month of his inaugural address at the
Prussian Academy of Sciences he found himself surrounded by a flood
of militaristic German nationalism, alone on a tiny island of pacifist
universalism. The Great War had begun, and his admired German
colleagues—Nernst, Haber, and even the mild-mannered Planck—
rushed to show their devotion to the fatherland. They and many other
academics signed an ill-conceived "Appeal to the Cultured World" de-
nying documented German war crimes in Belgium and asserting the
unity of German culture with German militarism. Einstein joined
with a physician friend, George Nicolai, to respond with a counter-
manifesto titled "Appeal to the Europeans" denouncing the attitude
of their colleagues as "unworthy of what until now the whole world
had understood by the word culture." No one of consequence agreed
to sign their version. For the next four years Einstein maintained his
critical stance toward both sides in the war but mainly expressed it, as
he put it, through the Socratic method—posing questions to his peers
that others did not dare raise. Because of his intellectual stature and

his identity as an outsider he was indulged as an eccentric genius, and suffered no major consequences (as he would later when anti-Semitism joined forces with postwar nationalism).

However, his colleagues did not confine themselves to written expressions of support for the war effort. Nernst, always a bit of a self-caricature, at age fifty volunteered as support staff in what was called the Drivers Corps. He drove off in his private car to the front just in time to participate in the first rapid advance on Paris, getting so far that he could see the glow of the city lights at night. When the drive stalled and trench warfare began, he returned to Berlin and devoted himself to developing nonlethal chemical weapons, whose prototypes failed to impress the General Staff. Haber had fewer scruples and focused on developing the truly deadly gas weapons for which the Germans became renowned: chlorine, mustard gas, and phosgene. If the General Staff had believed more strongly in these weapons, the Germans might well have broken through the Allied lines when they were first deployed at Ypres and won the war. Nernst and Planck would both lose sons in the war; Haber would lose his wife, Clara, who committed suicide at least partly out of disgust at her husband's embrace of deadly invention.[1] Remarkably Einstein never spoke out against his colleagues directly and actually defended them, saying that they were "internationally minded as scientists" in comparison to chauvinistic nonscientists.

While Einstein often mentioned the psychological toll of the war in his letters ("the international catastrophe weighs heavily on me"), he repeatedly demonstrated his preternatural ability to block out the external world and concentrate on his science. The years 1914 and 1915 were devoted to the completion of general relativity, the monumental achievement already discussed. Within months of that exclamation point, he had attacked the quantum problems with renewed vigor. Something big had happened in the atomic world during his cosmic interlude. The 1911 Solvay Congress, which had been of little

---

[1] Clara was reportedly particularly disturbed by the death of Haber's brilliant young collaborator, Otto Sackur, who was killed in a laboratory explosion while trying to improve gas weapons.

scientific use to Einstein, had been an eye-opener for a number of the distinguished participants from England and France, where the early quantum theory was barely known. Chief among them was a New Zealander, now transplanted to the physics laboratory at the University of Manchester in England, named Ernest Rutherford. He had been exploring experimentally the structure of the atom, blissfully unaware of the strong evidence from statistical physics that atoms obeyed new rules not explained by classical mechanics and electromagnetic theory. He had already been awarded the Nobel Prize in Chemistry for measuring the elementary unit of charge in the dramatic year of 1908, when Planck's Nobel nomination in physics was defeated owing to belated comprehension that his quantum of action, h, threatened the foundations of the field. By 1913 Rutherford's research group had done the decisive experiments supporting the idea of a nuclear atom: in this picture a number of light, negatively charged electrons orbited around a localized, much heavier nucleus with the same number of positive charges. Now one had a well-defined model within which classical physics could crash and burn; that was progress.

In Rutherford's lab at the time was a twenty-six-year-old visiting Danish physicist, Niels Bohr. This Bohr was showing distressing signs of preferring theory to experiment, not a popular trait in the laboratory of "Papa" Rutherford, but fortunately his manly vigor compensated for this weakness of character. Rutherford decided he was an acceptable breed of the species: "Bohr is different. He's a football player."[2]

Bohr in turn greatly admired Rutherford. After enduring an unsuccessful visit to J. J. Thomson in Cambridge before moving to Manchester, he wrote to his brother enthusiastically about the contrast. Rutherford's lab was "full of characters from all parts of the world working with joy under the energetic and inspiring influence of

---

[2] Football here refers to soccer of course. In fact Bohr was by all accounts a skilled goalkeeper, although not the athletic equal of his brother, Harald, who became a noted mathematician; Harald won a silver medal in soccer competing for Denmark in the 1908 Summer Olympics.

the 'great man.'" Rutherford himself was "a man you can rely on; he comes regularly and enquires how things are going and talks about the smallest details." Rutherford communicated to Bohr the extraordinary puzzles posed by the early quantum concepts. As Bohr recalled, "I got a vivid account of the discussions at the first Solvay meeting from Rutherford in 1911, shortly after his return from Brussels." He immediately set about finding a way to fit Planck's constant, h, into the Rutherford atom.

Bohr began his theorizing at just the right time. For some time the atom had been known to be an electrically neutral object containing negative electrons, which had been discovered by J. J. Thomson in 1897. These electrons appeared to inhabit a region of dimensions one hundredth of a millionth of a centimeter (10^{-8} cm). The nature of the positive charge, which counterbalanced the electron's negative charge and made the atom neutral, was not known in the first decade of the twentieth century. Elaborating on a proposal by Lord Kelvin, J. J. Thomson had introduced a "plum pudding" model in which the positive charge was spread out uniformly in a sphere of roughly the same dimensions as the electrons inhabited, that is, 10^{-8} cm. The electrons were somehow suspended in stationary rings within the positively charged sphere and were the "raisins" in the pudding. Classical physics required that the electrons be stationary (except when they were disturbed by some external perturbation) because, according to Maxwell's equations, electrons orbiting in the atom would continuously radiate their energy away, a process that was not observed. In contrast, isolated atoms in a gas emitted light only at a few narrow frequencies, specific for each element. Moreover they did not do this continuously but only intermittently or under external disturbance.

Thomson had not attended the Solvay Congress of 1911, where the necessity of changing fundamental laws at the atomic scale was communicated to Rutherford and others, and when he did attend the second conference, in 1913, he was still defending his classical atom. The distinguished continental attendees, who had been struggling with quantum paradoxes for many years, did not sit still for this. Lorentz interrupted his lecture to state, "The proposed hypothesis gives rise to an . . . objection of a general . . . nature. We can take it as established that a model in which everything happens according to the laws of ordinary mechanics will lead to Lord Rayleigh's formula for the black-body radiation [the ultraviolet catastrophe]. As the model proposed by Sir J. J. Thomson contains nothing which is incompatible with the rules of mechanics, it is highly improbable that we will be able to deduce from it the correct laws of radiation."

While Thomson was receiving this rude awakening, his former protégé Rutherford had already established experimentally that Thomson's plum pudding was indigestible. In radioactive decay, doubly ionized helium atoms, known as alpha particles, are ejected from the decaying atom; these can be collimated into a beam that can probe the structure of the atom. Starting in 1909, experiments by Hans Geiger[3] and Ernest Marsden, under Rutherford's supervision, found that alpha particles in such a beam were occasionally knocked completely sideways by a thin foil of metal atoms. As one might imagine from the pudding analogy, a diffuse sphere of positive charge as envisioned by Thomson could never result in such violent redirection of the alpha particle. As Rutherford put it, it was as if one had fired an artillery shell at tissue paper and found it ricocheted back at oneself. Thus there had to be something hard at the center of the atom, and since the electrons were known to be light, it could only be a highly localized and relatively heavy ball of positive charge, which he called the atomic nucleus. (Except for hydrogen, atomic nuclei contain equally heavy neutral particles, neutrons, with no charge, but that fact would not be established for two more decades.)

Thus when Bohr sat down to explain the mechanics of the atom in 1913, he knew several very important things. One could try to picture the atom as a miniature solar system, with light, negatively charged electrons orbiting around a localized, heavy, positively charged nucleus. The electrons had to orbit, because if they were ever stationary they would be attracted directly into the nucleus. Yet they couldn't radiate energy while they orbited, as required by classical electrodynamics;

---

[3] Inventor of the eponymous radiation counter.

otherwise they would eventually collapse into the nucleus anyway. But that was not a problem; that was a feature. He had grasped Lorentz's point, that his theory could only be successful if it *violated* some aspect of classical physics, and if it somehow got Planck's constant, h, into the equations. Blackbody theory had succeeded by using classical physics where it worked and changing the rules where it didn't. Why not take the same approach to the atom?

Bohr wrote his seminal paper on atomic spectra in early 1913, having returned to Denmark from England, and he sent it to Rutherford for his criticism. Rutherford was mightily impressed by the conclusions but objected to its length, as well as its wordy and ponderous style. "I really think you should abbreviate some of the discussions to bring it into a more reasonable compass," he wrote; "as you know, it is the custom in England to put things very shortly and tersely in contrast to the German method, where it appears to be a virtue to be as long-winded as possible." Far from being cowed by this reaction, Bohr traveled to Manchester and fought successfully for every single word of his manuscript. This semantically rich but equation-poor scientific writing style would be a lifelong characteristic of the Dane.

Bohr begins by noting that while his mentor, Rutherford, has established experimentally the plausibility of a "solar system" model of the atom, "in an attempt to explain some of the properties of matter on the basis of this atom-model we meet . . . with difficulties of a serious nature arising from the apparent instability of the system of electrons [due to radiative energy loss]. . . . The way of considering the problem has, however, undergone essential alterations in recent years owing to the development of the theory of energy radiation, and . . . affirmation of the new assumptions introduced in this theory, found by experiments . . . such as specific heat, photoelectric effect, Röntgen rays [x-rays], etc. The result . . . seems to be a general acknowledgement of the inadequacy of the classical electrodynamics in describing the behavior of systems of atomic size." Here, after mentioning, Einstein's two great predictions of experimental quantum behavior, the photoelectric effect and specific heat, he cites the proceedings of the First Solvay Congress. He continues, "Whatever the alteration of the laws of motion of the electrons may be, it seems necessary to introduce into [them] a quantity foreign to the classical electrodynamics, i.e. Planck's constant."

He realizes he is treading in Einstein's footsteps. "The general importance of Planck's theory for the discussion of the behavior of atomic systems was originally pointed out by Einstein [he cites Einstein's 1905 and 1906 papers on light quanta, and his 1907 paper on specific heats]. The considerations of Einstein have been developed and applied to a number of different phenomena, especially by Stark, Nernst, and Sommerfeld." Note that, as with Nernst in 1910, Einstein, not Planck, is seen as the promulgator of the quantum atom. But unlike Einstein in the years 1908–1911, Bohr will not try to put the constant h into Maxwell's equations. He will introduce it into classical mechanics by means of two stand-alone postulates, in a move very similar to Planck's original restriction $\varepsilon = nh\upsilon$ (which in Einstein's hands became quantization of vibrational energy).

So what was the problem with "Plancking" electron motion in the atom anyway? The problem is expressed by Einstein in his first series of letters to Lorentz: "For me the only difficulty consists in the fact that Planck's foundation (the introduction of the $h\upsilon$-quanta) does not apply to the elementary foundation of the theory, but only to the special case of oscillating structures [with a single frequency]. We do not know therefore, and cannot deduce . . . what kind of electrical and mechanical laws we have to introduce for the free electron . . . there is no contradiction here, but only the difficulty of generalizing Planck's approach." As already noted, Planck had taken the simplest possible mechanical system, a mass on a spring (what physicists call a linear oscillator) and appended the ad hoc rule that its energy could only be an integer times $h\upsilon$, a quantization rule with no justification in Newtonian physics.

The reason that a linear oscillator is so simple is that it makes a periodic motion (going back and forth on exactly the same path) *and* the frequency of this motion is independent of how far you stretch the spring. As discussed earlier, when you stretch the spring farther, you add more total energy to the motion; the extra energy exactly compensates for the extra distance it must travel in one period, so it oscillates over the longer distance in exactly the same time, that is, with the same frequency. So for a linear oscillator the frequency of oscillation is a constant, independent of the energy of the oscillator, and the rule $\varepsilon = nh\upsilon$ determines what energies are allowed. When a Planck oscillator emits a light quantum, $h\upsilon$, it loses one quantum of energy, now having $\varepsilon = (n - 1)h\upsilon$, but the frequency of motion doesn't change, because it is a linear oscillator. The same frequency, υ, describes both the frequency of the electronic motion *and* the frequency of the emitted light, just as one would expect in classical electrodynamics: oscillating charges produce radiation at their oscillation frequency.

But this single-frequency behavior of the linear oscillator is due to its simple force law; the restoring force of the spring (which pulls it back and forth) is proportional to the distance the spring is stretched. The force on an electron in a nuclear atom is not of that kind; Bohr assumed (correctly) that it was the inverse square law of electrical attraction between oppositely charged particles. At least for the simplest case, the hydrogen atom (one electron moving around one proton), this force law leads to periodic orbital motion,[4] but the frequency of the motion changes continuously as the energy is changed. For example, two electrons in circular orbits of different orbital radii not only have different "binding energies" to the nucleus but also have different orbital frequencies. Since the math involved is exactly the same as for a planet orbiting the Sun, all the equations for how this works were well known from classical celestial mechanics. In particular, planets farther away from the Sun are less strongly attracted to it, have less binding energy, and hence orbit the Sun with a longer period (lower frequency); Pluto's year, for instance, is 248 Earth years. Bohr was prepared to use the same principles of classical mechanics for electron orbits; but then how does one implement a quantum rule for the allowed energies when, unlike for an oscillator, the frequency changes when the energy does?

---

[4] As long as the electron is bound to the nucleus (i.e., doesn't have enough energy to escape from the nuclear attraction).

Let's call the electron orbital frequencies f to distinguish them from the frequencies of light the atom might emit, which we will still call v. By a rather tenuous argument, Bohr came up with a surprising answer. The allowed energies for electrons orbiting the nucleus were given by a whole number times Planck's constant times *half* the final orbital frequency.[5] Planck's $\varepsilon = nhf$ for oscillators became Bohr's $\varepsilon = nhf/2$, for electron orbits in hydrogen. The unique circular orbits picked out by this rule were termed "stationary states," which were postulated to remain stable forever, without radiating energy. Since, as noted, the frequencies of the orbits themselves vary with the energy of orbit, this rule does not lead to equally spaced energy levels, in contrast to what Planck had found for oscillators.

In coming up with this less-than-obvious generalization of Planck's rule, Bohr was guided by remarkable intuition, and by a critical observation about the spectra of light emitted from hydrogen gas. It had been known for a very long time that the light emitted from atomic gases does not consist of a continuous "rainbow" of colors but instead has a few very pure colors. When this light was focused onto an optical device called a diffraction grating, which functions essentially like a prism, the different colors could be separated from one another in space and then projected onto a screen, giving rise to narrow bands, or "spectral lines," of color. Since different elements give different sequences of colors (lines), one could use such spectra to identify which elements were emitting the light. This was, of course, of great importance for disciplines such as physical chemistry and also for astronomy, where it could be used to identify the elements present in distant stars. But why certain lines appeared for a certain element, and why the lines were spaced in certain patterns, were completely mysterious from the point of view of fundamental physics.

[5] Bohr got this answer by considering the average frequency of all circular orbits starting infinitely far away from the nucleus, where $f = 0$, all the way in to the final orbit, with frequency f. Bohr's restriction can be alternatively phrased as the constraint that the angular momentum of an electron in a Bohr orbit is equal to $nh/2\pi$, which is the form that generalizes to other force laws and is emphasized in modern accounts. This idea was expressed but not strongly emphasized in his original paper.

In 1888 the Swedish physicist Johannes Rydberg had noted that the frequencies of hydrogen spectral lines seemed particularly simple: they formed a series that could be generated by a simple arithmetic formula.[6] Bohr realized that the specific generalization of Planck's rule that he had come up with, combined with standard classical formulas relating the frequency of an orbit to its energy, would give rise to Rydberg's spectral series. That looked extremely promising.

But he had to add one more critical step to explain Rydberg's observations. Rydberg did not directly observe *electron* energies; he observed the frequencies of the *light* emitted from the atom, presumably as a consequence of the change of energy state of the orbiting electron. Bohr postulated that when the electron absorbed or emitted energy by changing its orbital state, the frequency, v, of the light that was emitted or absorbed was determined by the Planck/Einstein rule: $\varepsilon_1 - \varepsilon_2 = hv$, where ε_1 is the initial energy of the electron and ε_2 is its final energy. This assumption, combined with his rule for quantizing electron energies described above, implied that only certain frequencies of light could be emitted or absorbed by hydrogen atoms, exactly those observed by Rydberg!

That was the good news. There was also bad news. With this argument Bohr did something so radical that even Einstein, the Swabian rebel, had found it inconceivable; Bohr *disassociated* the frequency of the light emitted by the atom from the frequency at which the electron orbited the atom. In the Bohr formula, $\varepsilon_1 - \varepsilon_2 = hv$, there are two electron frequencies, that of the electron in its initial orbit and that of the electron in its final orbit; *neither* of these frequencies coincides with the frequency, v, of the emitted radiation! This was a pretty crazy notion to a classical physicist, for whom light was *created* by the acceleration of charges and must necessarily mirror the frequency of the charge motion. Bohr admitted as much: "How much the above interpretation differs from an interpretation based on the ordinary electrodynamics is perhaps most clearly shown by the fact that we have been forced to

[6] The ratios of the distances in frequency between pairs of lines corresponded to the difference of the inverse of the squares of integers.

assume that a system of electrons will absorb radiation of a frequency different from the frequency of vibration of electrons calculated in the ordinary way." However, he noted, using his new rule, "obviously, we get in this way the same expression for the kinetic energy of an electron ejected from an atom by photo-electron effect as that deduced by Einstein." So, as a final justification, he relied on exactly the experimental evidence that motivated Einstein's light-quantum hypothesis and inaugurated the search for a new atomic mechanics.

But Bohr's atomic theory was hardly the new mechanics for which Einstein had been searching. There was still no underlying principle to replace classical mechanics, just another ad hoc restriction on classical orbits, a variant on Planck's desperate hypothesis. So be it, Bohr decided; his gambit was not elegant, but it worked! Not only did it explain the hydrogen spectrum for visible light, which consisted of two different series named for their discoverers, the Balmer and Paschen series; it also predicted a series of ultraviolet lines that had not been observed. Unknown to Bohr, the American physicist Theodore Lyman had already begun detecting this series during the preceding few years, and his results, shortly to be published, would be found to agree with the Bohr formula. But the pièce de résistance was the application of Bohr's formula to puzzling astrophysical spectral lines. A new series of spectral lines had been observed in the blue supergiant star ζ-Puppis by the astrophysicist E. C. Pickering and subsequently produced on earth, in helium-hydrogen mixtures, by another astrophysicist, Alfred Fowler. These lines corresponded to every *other* line in the Balmer series for hydrogen and, for want of a better explanation, had been tentatively assigned to hydrogen. Bohr realized that these lines were explained perfectly by his formula. All he had to do was assume that they arose from singly ionized helium atoms in the star's photosphere. Singly ionized helium has only a single electron, like hydrogen, but of course it has a larger nuclear charge and mass; simply plugging into Bohr's formula the new values of charge and mass gave the observed spectral lines.

Bohr's explanation of the Pickering-Fowler spectrum was the clincher; the theory wasn't just a tautology designed to fit the known

FIGURE 20.1. Albert Einstein and Niels Bohr in discussion, circa 1925–30. Photograph by Paul Ehrenfest, courtesy AIP Emilio Segrè Visual Archives.

hydrogen spectrum. It had predictive and explanatory power. When he became aware of this, Einstein was mightily impressed. Einstein's old Zurich acquaintance, the chemist George Hevesy, was the first to tell him the news. He met Einstein at the German Physical Society conference in Vienna in 1913 and, as he wrote Rutherford, "we came to speak of Bohr's theory, he told me that he had once similar ideas, but did not dare to publish them. 'Should Bohr's theory be right, it is of the greatest importance' [he said]. When I told him about the Fowler spectrum the big eyes of Einstein looked even bigger and he told me, 'Then it is one of the greatest discoveries.'" In writing to Bohr, Hevesy recounted a further highly revealing comment from Einstein: "I told him [the explanation of the Fowler spectrum]. . . . When he heard this he was extremely astonished and told me: '*Then the frequency of the light does not depend at all on the frequency of the electron* [italics added] . . . this is an enormous achievement. The theory of Bohr must then be right.'"

So this was the revolutionary step that Einstein, who had mused on spectral lines as early as 1905, had not been willing to take. The frequency of light had nothing to do with the frequency of motion of the electron in the atom—who would have guessed? Einstein, the originator of so many "crazy" leaps of intuition himself, could recognize one when he saw it. What impressed him were not Bohr's calculations, which were simple, but the insight to guess what one should keep and what one should drop from the laws of classical physics. Much later he lauded Bohr's achievement thus:

> All my attempts, however, to adapt the theoretical foundation of physics to this [quantum] knowledge failed completely. It was as if the ground had been pulled out from under one, with no firm foundation to be seen anywhere, upon which one could have built. That this insecure and contradictory foundation was sufficient to enable a man of Bohr's unique instinct and tact to discover the major laws of the spectral lines and of the electron-shells of the atoms, together with their significance for chemistry, appeared to me like a miracle—and appears to me as a miracle even today. This is the highest form of musicality in the sphere of thought.

Inspired by the establishment of the Bohr postulates for the atom, Einstein was able to write a crucial coda to Bohr's composition shortly after he returned to quantum cogitation in 1916, one that would add a new framework to the insecure foundation of the developing quantum theory.

CHAPTER 21

RELYING ON CHANCE

Fundamental as [the relativity theory] of Einstein has proved to be for the development of the principles of physics, its applications are for the present still at the very limit of the measurable. His tackling of other questions that are at the moment at the center of attention has proved to be much more significant for applied physics. Thus he was the first to demonstrate the significance of the quantum hypothesis also for the energy of atomic and molecular motions. . . . That he might sometimes have overshot the target in his speculations, for example in his light quantum hypothesis, should not be counted against him too much.

—PLANCK, NERNST, RUBENS, AND WARBURG, LETTER
NOMINATING EINSTEIN FOR THE PRUSSIAN ACADEMY
OF SCIENCES, JUNE 12, 1913

"Why Planck and I engaged him just as you take on a butler, and now look what a mess he's made of physics; one can't turn one's back for a minute." This was Nernst's sardonic appraisal of Einstein's triumph with general relativity. The esteemed professors of the Prussian Academy had not brought Einstein to Berlin to monkey around with the law of gravity and the geometry of space-time; they were expecting him to lead them to victory in the race to understand the atom. Instead he had ceded the inside rail to the Dane Niels Bohr, and his English collaborators, just as the Great War had begun. Moreover they, along with most of the physics community, had never accepted Einstein's notion of a new theory in which wave-particle duality would naturally emerge and light would really have particulate properties.

One might have thought that Bohr's theory would have advanced the belief in photons. Bohr's second postulate stated that when an electron makes a transition between two of its allowed orbits, the result is "the emission of homogeneous radiation, for which the relation between the frequency and the amount of energy emitted is the one given by Planck's theory [$\varepsilon = h\upsilon$]." If one squinted just a little at that sentence (as indeed one often had to do with Bohr's writing), it sure seemed like he was saying that the atom emitted one Einsteinian light quantum every time an electron changed its energy state. But Bohr did *not* mean to say this and in fact remained opposed to the particulate concept of light for the next twelve years. He believed that quantum mechanics governed the atom but that the light emitted in the transition between allowed orbits, while it contained the fixed quantum of energy, $h\upsilon$, was nonetheless solely an electromagnetic wave.

Einstein himself had been chastened by his failure to find quanta in a modified Maxwellian electrodynamics and since 1911 had spoken of photons only sotto voce. In May of 1912 he expressed his position to Wien: "One cannot seriously believe in the existence of countable quanta, since the interference properties of light emitted by a luminous point in different directions are not compatible with it. Nevertheless, I still prefer the "honest" theory of quanta to the hitherto found compromises meant as its replacements." Einstein here identifies a puzzle that he had been struggling with for several years already in 1912. In classical electromagnetism a point source of radiation (such as an atom) emits a uniform spherical wave of light traveling outward in all directions; nonetheless if one refocuses some of these diverging rays to a screen, they will show standard interference patterns. Einstein and Lorentz had already agreed that each single light quantum was capable of interfering with itself, so such a picture suggested that light quanta from a point source must "break up" in some manner, belying their existence as "countable quanta." Another issue Einstein had pondered was the phenomenon of radioactivity, in which one or more particles is emitted from the nucleus, apparently at a random time and in a random direction. Already in 1911 he saw a parallel here with electromagnetic radiation, writing to Besso, "the process of absorption [of light] . . . really does have similarity with a radioactive process." His close friend and future Nobel laureate Otto Stern recalls how much this problem vexed him during his Prague years: "Einstein always wracked his brain about the law of radioactive decay. He constructed such models." After Bohr's eye-opening atomic theory, it struck him that the new picture of emission and absorption could permit these troublesome puzzle pieces to fit together.

In February of 1916 Einstein had already put general relativity aside and begun to catch up on the quantum theory of atoms. Sommerfeld had written Einstein in December of 1915 to ask him to look at an improvement he had made in Bohr's formulas that Einstein would find particularly interesting. Sommerfeld had realized that there was no need for Bohr to restrict the electrons to circular orbits; they could also move on elliptical orbits (as do all the planets[1] in our solar system). He then used a variant on Bohr's approach, which he had invented, to determine the quantized energies of the allowed orbits.[2] He found a more general formula than Bohr's that initially gave the same results for the hydrogen spectra. But in his new approach he was able to take into account an additional effect that Bohr could not. Einstein's special relativity theory predicts that the measured mass of an electron will increase with increasing velocity. Electrons whizzing around the nucleus were calculated to be moving at a significant fraction of the speed of light, and so this increase in mass should have a measurable effect on the electron's orbital frequency. Including this effect, as Sommerfeld now had done, caused the spectral lines in the hydrogen series to split into closely spaced groups of lines ("fine structures"), their number depending on the final state of the electron after light has been emitted. Sommerfeld wondered in his letter to Einstein whether the newly minted general theory of relativity would affect

---

[1] Circular orbits are allowed by classical mechanics but require a specific relationship between a planet's orbital energy and its angular momentum, which never is precisely satisfied when a planet forms out of primordial matter. In our solar system, however, planetary orbits are quite close to being circular.

[2] This approach is now called Bohr-Sommerfeld-Wilson quantization; it will be discussed further below.

his calculations, but Einstein assured him that these effects were too small to matter in this context. The "fine-structure" effect had been previously seen and, just at that time, had been measured carefully by the noted German spectroscopist Friedrich Paschen (after whom one of the original hydrogen series was named). In late December of 1915 he wrote to Sommerfeld, "My measurements . . . agree everywhere most beautifully with your fine structures." The experimentalist was so delighted by this transformation of an experimental anomaly into an important discovery that he is reported to have loudly exclaimed, "Now I believe in relativity theory!"

It was this beautiful marriage of Bohr's atomic quantum theory and relativity principles that so impressed Einstein, once he had digested it; this is the work that in February of 1916 he called "a revelation" in an ecstatic letter to Sommerfeld. Later, in August, in the midst of his own work, inspired by Bohr's theory, he wrote Sommerfeld again to say, "your spectral analyses are among my finest experiences in physics. It is just through them that Bohr's idea becomes entirely convincing. If only I knew which little screws the Lord is adjusting there!" By then he had already shown, in a paper submitted on July 17, that Bohr's single postulate, that electrons make transitions between allowed stationary energy states via absorption and emission of radiation of energy $h\upsilon$, had remarkable implications.

In this first paper of 1916, titled "Emission and Absorption of Radiation in Quantum Theory," he returns to the theme he first expounded in 1909, that Planck's derivation of the blackbody law is contradictory because it *uses* classical electrodynamics to relate the mean energy of an oscillator to the energy density of the radiation field but then *departs* from classical physics to calculate this mean energy according to a quantum prescription. He again praises Planck's courage to leap into the unknown—"his derivation was of unparalleled boldness—but adds, "however it remains unsatisfactory that the electromagnetic-mechanical analysis [used] is incompatible with quantum theory." He continues: "Since Bohr's theory of spectra has achieved its great successes, it seems no longer doubtful that the basic idea of quantum theory must be maintained." In the interests of consistency, he says, Planck's classical assumptions must be "replaced by quantum-theoretical contemplations on the interaction between matter and radiation. In this endeavor I feel galvanized by the following consideration, which is attractive both for its simplicity and generality."

The simple, general consideration he mentions flows from the concept of thermal equilibrium, which we have encountered earlier. The blackbody law holds for radiation in contact with matter, so that the entire system (radiation + matter) has settled into the most probable thermodynamic state. In the thermal equilibrium state the entropy of the system is at its maximum value, its temperature is no longer changing, and the average energy of both the radiation and the matter is not changing in time.[3] But since the matter and radiation are continually exchanging energy, this state of equilibrium is not to be conceived of as the absence of interaction but rather as continually compensating change. One can imagine the two systems (matter and radiation) as two swimming pools connected by pipes, so that water is flowing from one pool into the other through certain pipes and being pumped back into the first pool through other pipes, and on average the level of water [energy] in each pool does not change.

Following Bohr, Einstein can now be more precise about the nature of this dynamical equilibrium state. He assumes the matter in contact with radiation is a gas of molecules with discrete, quantized energy levels, Bohr's stationary states. His argument is, however, so general that he never needs to assign specific values to the allowed energies: it is enough that they exist. Thus his reasoning becomes independent of the details of Bohr's method for calculating atomic energy levels. Energy is then exchanged between the molecules and the radiation field according to the Bohr prescription. If a molecule has any energy greater than its minimum energy state (called the "ground state"), then it can either emit radiation of the appropriate energy, hv_1, so as to drop down to a lower

---

[3] Recall that Jeans's discredited explanation for the blackbody radiation observations was based on the hypothesis that matter and thermal radiation were not in thermal equilibrium at high frequencies. Now there was general agreement that this was incorrect, and Einstein could base his new work on the assumption of equilibrium without fear of such criticism.

energy state, or absorb radiation energy of another specific frequency, $h\nu_2$, and jump up to a higher energy state. In order for equilibrium to be maintained, these processes of emission and absorption of energy from the radiation must balance out on average; the same amount of energy must flow into the molecule as out. In fact, Einstein points out, not only must the energy flow balance out overall; the exchange of energy must balance out independently for *each pair* of molecular levels. For each such pair "upward transitions" (absorption) must equal "downward transitions" (emission) or the system could not maintain its equilibrium.

So far so good; the logic and math are so simple he can hardly have made a mistake, but he also can hardly have gained much deductive power. Then, a crucial insight: "We shall distinguish here two types of transitions." When the molecule is in any of its higher-energy, "excited" states, *even if there is no radiation at all present*, he assumes that there is still some chance that it will emit radiation "without external influence. One can hardly imagine it to be other than similar to radioactive reactions." What a leap! Radioactive decay was a mysterious *nuclear* phenomenon that appeared to be completely random. A radioactive substance has a certain half-life, that is, the period of time during which on average half of the nuclei emit radioactive particles. But for any specific nucleus all one could say was that the probability was one-half that it would not decay in one half-life and one-fourth that it wouldn't decay in two, etcetera. The actual time and direction of decay was (and is still) unpredictable. Now Einstein is claiming that at least *some* of the events in which atoms emit light are just like this, are what we now call "spontaneous emission" events, and he writes down exactly the same mathematical rule for the number of spontaneous emission events per unit time as for radioactive decay.

He then considers other emission events of a more conventional sort, which we now call "stimulated emission"; the number of *these* events depends on the amount of radiation that is already present, which means that in equilibrium this number will be proportional to the blackbody radiation density. These events would have been more familiar to his readers because in classical electrodynamics one pictures the radiation field as "driving" or "being driven by" the electron charges in the atom, increasing or decreasing the amount of energy contained in the electron's orbital motion. When the radiation field subtracts energy from the electrons, we have "stimulated emission"; when it adds energy to the electrons, we have absorption. Whew, at least some of this sounds familiar, but . . . he completely throws out the classical method of calculating absorption and emission. *Instead he treats* these *processes as random too.*

Now all he has to do is balance out the various processes; the energy lost by the molecules in spontaneous and stimulated emission must on average equal that gained by absorption. By one last sleight of hand he relates the rate of stimulated emission to that of absorption, simplifying the equation of balance. Two more lines of algebra and, *Mein Gott,* he has derived Planck's radiation law! Einstein has ingeniously bypassed all the complicated counting and gone straight to the answer. He can't resist giving himself a little pat on the back: "The simplicity of the hypotheses, the generality with which the analysis can be carried out so effortlessly, and the natural connection to Planck's linear oscillator . . . seem to make it highly probable that these are the basic traits of a future theoretical representation." In his jubilant letter to Besso a few weeks later he is even more effusive: "A brilliant idea has dawned on me about radiation absorption and emission. . . . An astonishingly simple derivation, I should say *the* derivation of Planck's formula. A thoroughly quantized affair."

Einstein was correct; his approach has become *the* derivation of Planck's formula. It is completely valid within the modern theory of quantum mechanics and electrodynamics and is in fact the reasoning still used in the majority of textbooks. Einstein introduced two unknown constants of proportionality (one for the rate of spontaneous emission, denoted by A, and one for the rate of stimulated emission, denoted by B), and then used additional arguments to replace them with known constants. These new fundamental quantities can be calculated directly in the modern theory, but in homage to the master are still labeled the "Einstein A and B coefficients."

Einstein did not, however, rest on his laurels. Rederiving Planck's law in a purely quantum framework was progress, but it did not in itself clarify the basic question of the existence of light quanta. He was still hunting for the resolution of the "spherical wave paradox," that a classical point source emits a uniformly expanding spherical wave front, like the ripples a rock makes when dropped into a pond. Such waves seemed to rule out conceptualizing atomic emission as the release of a localized particle of light, which flies out in a particular direction. It was hard to visualize dropping a rock in a pond and having a single bump of water move out in a specific direction. However, it wasn't obvious that single atoms really behaved like classical point sources; maybe, he thought, the classical viewpoint was just wrong. Perhaps real atomic emission was a directed process, emitting "bumps" of light in specific directions; only when there were many atoms randomly emitting in all directions (or one emitting repeatedly) would it *appear* spherical.

A few weeks after his first 1916 paper Einstein realized, to his delight, that his new hypotheses of quantum emission allowed him to prove just this fact, and he eagerly drafted a second paper containing this demonstration. When he wrote to Besso on August 11, crowing about having found "*the* derivation," he added, "I am writing the paper[4] right now." In a follow-up letter to Besso two weeks later he added, "it can be demonstrated convincingly that the elementary processes of emission and absorption are directed processes."

So what was the new insight that so excited him? Einstein had realized that he could return yet again to the well that had yielded the theory of Brownian motion of particles in suspension and of radiation energy fluctuations. In this famous second quantum paper of 1916, titled "On the Quantum Theory of Radiation," he reviews his elegant new derivation of Planck's law, stating, "this derivation deserves attention not only because of its simplicity, but especially because it seems to clarify somewhat the still unclear process of emission and absorption of radiation by matter." Now, he says, we must go beyond just considerations of energy exchange. "The question arises: does the molecule

---

[4] This second paper did not become available until 1917 and is the one usually cited and discussed, so it is not widely appreciated that the key ideas were found between May and August of 1916, only six to nine months after the completion of general relativity.

receive an *impulse* (i.e. a push in a specific direction) when it absorbs or emits the energy, *hv*? . . . *It turns out that we arrive at a theory which is free of contradictions, only if we interpret those elementary processes as completely directed processes* [italics in the original]. Herein lies the main result of the following considerations."

We have already seen that radiation exerts pressure even in classical electrodynamics and thus can push on matter, that is, transfer momentum to matter. Einstein had used this fact in his Salzburg "sliding mirror" thought experiment. The effect is similar to the recoil that occurs if you fire a gun; a light wave emitted in a specific direction causes the emitting atom to recoil in the opposite direction. Similarly, the analogue of absorption of light waves by an atom is the unfortunate process of "absorbing" an incoming bullet, which among the less problematic of its effects causes the "absorber" to be pushed in the direction of motion of the bullet. However, if the atom emitted a spherical light wave, the recoil pressure would be equal in all directions and no net momentum would be transferred. One can picture this by imagining a platoon of soldiers on a raft. If they all line up and fire their rifles in the same direction, then the recoil pushes the raft strongly in the opposite direction, but if they form a circle and fire outward at the same time, the raft will not move. (They could also form a circle and all fire inward, and again the raft would not move, but that would have other effects.) So the consequences of directed as opposed to undirected (spherical) emission are different, and in the process of energy exchange between molecules and radiation that Einstein is discussing, different motions of the molecules occur, depending on whether one assumes the emission and absorption is directed or undirected ("isotropic").

To prove that each molecular interaction with radiation is a directed process, he makes the following argument. Imagine a gas molecule moving around in an enclosure filled with radiation, both of which are at the same temperature, T (this is the condition for thermal equilibrium). For the radiation this means that its energy density will be described by a universal radiation law, which depends on T and on the frequency, v. Einstein is not here assuming that this law is given by the (now) well-known Planck formula; his goal here is to derive the Planck law in a new way, based on Bohr's quantum atom and general considerations from statistical physics. In this effort he assumes that the atoms in the gas have a kinetic energy given by our old friend the equipartition theorem, which he knew failed for the vibrational energy of molecules but was well confirmed for the energy associated with the free motion of atoms in a gas.[5]

So our equally partitioned gas molecules are gliding along with the average kinetic energy $3kT/2$, but their motion is not free of all forces. The presence of all that radiation generates a kind of frictional force, due to the Doppler effect. The Doppler effect, which is familiar for sound waves, is the observed change in frequency (pitch) that occurs when a sound source is moving toward or away from the receiver: when moving toward, its pitch is measured to be higher than when at rest, and similarly it is found to be lower when the sound source is moving away. The same effect occurs for light waves: when you move toward them their frequency increases, and when away from them it decreases. Actually the details are a bit different for light because, unlike sound, it always is measured to move at c, but Einstein had worked out the formulas for this while banishing the ether way back in the miracle year of 1905.

Why does this effect cause friction? Imagine a situation where water waves of equal frequency are being generated in opposite directions at two ends of a swimming pool and you are standing in the middle, being buffeted forward and backward by those waves from each side. If you stand still, then on average you get an equal number of forward and backward shoves as the waves hit you, and you are not, on average, pushed toward either end of the pool. However, if you move at some reasonable speed (compared with the speed of the waves) toward one end, then you hit the incoming waves from that end more frequently than the waves generated behind you at the opposite end. This is the

[5] Eight years later Einstein would be the first to discover that the equipartition theorem can break down even for an atomic gas (see chapter 25), but those effects require such low temperatures that they would not become observable until the end of the twentieth century. Moreover, this fact does not invalidate the argument he is making in the current work.

Doppler effect in a very concrete form: you are moving in a medium in which waves are also moving, such that you encounter more crests (at a higher frequency) when you move against the direction of wave propagation, and at a lower frequency when you move in the same direction as the waves. In this case the visceral effect is that you get knocked backward more than you get knocked forward, and you feel an effective force impeding your motion toward the end of the pool you are approaching. But if you turn around and start walking toward the other end, exactly the same thing happens, except now the force is pointing in the opposite direction; that is, it behaves like friction, slowing you down no matter which direction you go. It takes Einstein a dense three pages of algebra to work out the exact mathematical formula for this frictional force, but this is the essential idea.[6]

But this is not the only force acting on the molecule; it cannot be, because if it were, over time the radiation field would extract all the kinetic energy from the molecules, leaving them at absolute zero temperature. (In our pool analogy, the walker gets too tired to walk against the current and just stands still.) Again we would have a version of the ultraviolet catastrophe. But Einstein knows how nature avoids this. The previous reasoning assumed that the absorption events occur in a perfectly regular sequence, whereas in actuality the molecule is being randomly buffeted by photons at irregular intervals, so that in any short interval it gets a net kick from the radiation that can push it in either direction, forward or backward. Einstein calculates the magnitude of this *fluctuating* force. And then he assumes that these two forces, the frictional one and the fluctuating one, must on average balance, precisely to avert the unobserved cooling of matter by radiation. But this balance equation depends on the mathematical form of the radiation distribution law, the infamous universal function $\rho(v, T)$. With great relish, Einstein shows that Planck's law, and only Planck's law, will make the two forces cancel each other on average.

---

[6] The analogy here is not perfect, because there is no ether in which light waves move, but as noted, there is a relativistic version of the Doppler effect which still leads to a frictional force on the gas.

But central to all of Einstein's reasoning is that each emission and absorption event is a *directed* process. "If we were to modify one of our postulates about momenta [forces], a violation of the [force balance] equation would be the consequence. . . . To agree with this equation— which is demanded by the theory of heat—in a way other than by our assumptions seems hardly possible." He concludes, "If a molecule suffers a loss of energy in the amount $h\upsilon$ . . . then this process is a directional one. There is no emission of radiation in the form of spherical waves."

Not only has Einstein resolved a paradox in his own mind; he also has changed the nature of the evolving quantum theory. Werner Heisenberg, one of the founders of modern quantum mechanics, has pointed out that "[Einstein] himself, in his paper of [1917], . . . introduced such statistical concepts [into quantum theory]." Pascual Jordan, a key collaborator of Heisenberg's, described Einstein's paper as among the most important to influence the development of modern physics. From that point on, random, acausal processes would be integral to the theory. This was not a concept contained in the Bohr-Sommerfeld theory of the atom; it was Einstein who let this unwelcome genie out of the bottle. He would come to regret it.

He continues: "the molecule suffers a recoil . . . during this elementary process of emission of radiation; the direction of the recoil is, in the present state of theory, determined by 'chance' . . . the establishment of a quantumlike theory of radiation [appears] almost unavoidable. The weakness of the theory is . . . that it does not bring us closer to a link-up with the wave theory . . . [and] also leaves the time of occurrence and direction of the elementary processes a matter of 'chance.' Nevertheless, *I fully trust in the reliability of the road taken* [italics added]."

Einstein was confident of his results not just because of the simplicity and elegance of the logic; he now believed he had attained the long-sought proof that light quanta were as "real" as any other elementary particles, not just a manner of speaking about the interaction of radiation with matter, as maintained by Planck, Lorentz, and others. He proclaimed as much in his next letter to Besso: "any such elementary process is an *entirely directed process*. Thus the light quanta are as good as established."

CHAPTER 22

CHAOTIC GHOSTS

"I have firmly decided to bite the dust with a minimum of medical assistance when my time has come, and up to then to sin to my wicked heart's desire. Diet: smoke like a chimney, work like a horse, eat without thinking and choosing, go for walks only in really pleasant company, and thus only rarely, unfortunately, sleep irregularly, etc." This was Einstein's cheeky pronouncement to Elsa Einstein back in August 1913, before his arrival in Berlin and the monumental labors that occupied him between then and the completion of his new work on thermal radiation in 1916. Many historians regard the period of November 1915 to February of 1917 as Einstein's second miraculous phase. During this period he produced fifteen papers, including the final form of the theory of general relativity, its first extensions into cosmology, as well as the next conceptual pillar in the emerging quantum theory, the ideas of spontaneous emission, intrinsic randomness, and the marriage of the Bohr atom with the Planck law, implying the reality of photons. And all this was accomplished in the midst of wartime and the steadily increasing hardship of daily life as hostilities dragged on and Germany's prospects dimmed. By early 1917 an exhausted and ill Einstein would have to reconsider how seriously he intended to ignore the demands of his body.

The winter of 1916, in which Einstein returned to quantum theory with renewed intensity, became known as the "turnip winter" in Berlin as the lowly turnip was fashioned into all manner of absent foodstuffs: bread, cake, coffee, and even something purporting to be "turnip beer." The British were blockading food shipments, and as a result during that year of 1916 an estimated 120,000 Germans died from malnutrition. In February of 1917 Einstein, along with the rest of Berlin, was suffering through an unusually frigid winter, during which he fell ill with liver and bladder ailments that reached life-threatening severity, causing him to lose over fifty pounds in two months. Einstein had not suffered major privations during these years, thanks to packages of supplies sent to him by his friend Zangger in Switzerland and his relatives in southern Germany, so his illness was due to mainly to overwork, poor eating habits, and a chronically troubling digestive system, which Mileva referred to as his "famous complaint." Having only just presented his new work on cosmology to the Prussian Academy on February 6, 1917, he took to his bed and on the fourteenth wrote to Paul Ehrenfest in Leiden canceling his planned visit to Holland. "I am quite infirm from a liver condition," he explained, "which imposes on me a very quiet lifestyle and the strictest diet and regimen." Two months later he wrote to Lorentz, "I have not been working much at all, and that under ideal circumstances." By May he was singing a different tune than the exuberant overture he had sent to Elsa four years earlier. He told Besso he had resisted the doctor's order to go for a "spa cure," saying he could not "raise the necessary superstition"; but, he continued, "I am committing myself to do everything else—which is unbelievable—to abstain from drinking, etc., in short to perform the rites of medicine loyally and piously."

Having failed in a first attempt to obtain a divorce from Mileva, Einstein had maintained until this period a certain distance from his cousin Elsa, no longer being so eager to jump into a second marriage as in the heady early days of 1913. In the midst of his acute illness he would still write to Zangger, "I have come to know the mutability of all human relationships and have learned how to insulate myself against heat and cold, so the temperature is quite steadily balanced." But Elsa now took the lead in nursing him back to health and regulating his convalescence; by the end of the summer of 1917 she had procured for him the apartment next to hers at Haberland Strasse 5 and had even moved his things into it while he was away traveling. By December he could report to Zangger, "my health is quite fair now. . . . I have gained four pounds since the summer, thanks to Elsa's good care. She cooks

FIGURE 22.1. Watercolor of Einstein and Paul Ehrenfest playing duets in Leiden during one of Einstein's periodic visits. Original watercolor by Maryke Kamerlingh-Onnes, courtesy AIP Emilio Segrè Visual Archives.

everything for me herself, as this has proved necessary." However, by January he was bedridden again for six weeks and did not feel fully healthy again until the following summer, despite "Elsa indefatigably cooking" his "chicken feed." It was in that summer that Einstein finally received the consent to a divorce from Mileva, with its famous stipulation that she would receive the proceeds of his inevitable Nobel Prize (should he survive long enough to receive it). By the following June (1919), after the legal formalities had been concluded, Einstein would finally marry Elsa and fulfill her long-held desire to become Mrs. Albert Einstein. She would be a steady, reliable presence in his life for the next two decades but never the romantic companion he had imagined in his early love letters, written before actually moving to Berlin.

Einstein's health problems, beginning in February 1917 and continuing well into 1918, along with the complex and draining personal issues of divorce and remarriage, made these two years less scientifically productive than the previous two had been. He continued to work on elaborations and popularizations of general relativity, but one senses that quantum theory and the new atomic mechanics remained paramount in his research ambitions. In March of 1917, while still too ill to do much, he wrote to Besso referring to his new paper on thermal radiation, which had only recently been published despite its provenance nine months earlier. "The quantum paper I sent out has led me back to the view of the spatially quantum-like nature of radiation energy. But I have the feeling that the actual crux of the problem posed to us by the eternal enigma-giver is not yet understood absolutely. Shall we live to see the redeeming idea?"

The very next day he wrote to Zangger bemoaning his health and his lack of intellectual momentum: "Scientific life has dozed off, more or less; nothing is going on in my head either. Relativity is complete, in principle, and as for the rest, the slightly modified saying applies: . . . what he can do he does not want; and what he wants he cannot do." Coming on the heels of his proclamation about the quantum enigma to Besso, his self-appraisal seems clear: what he wants most at the moment is to truly understand what is going on with atoms and their interaction with light.

Following his "perfectly quantic" derivation of the Planck law, Einstein's period of vacillation on the reality of light quanta, begun in 1911, was over. He was convinced that light quanta were full-fledged particles, which were localized in space, moved along directed trajectories, and carried momentum as well as energy. This conviction simply renewed the challenge of how to reconcile their reality with the interference properties exhibited by electromagnetic radiation, which seemed to require that light extend over large regions of space. While no idea had arisen, either from him or from the expanding community of quantum physicists, for a new mathematical theory of light that could encompass these two conflicting aspects, around this time Einstein began to develop a conceptual framework that could serve as a stopgap measure on the way to a fuller theory.

He hypothesized that light is emitted in a twofold process. While a guiding wave obeying the classical Maxwell equations is generated, at the same time some number of localized light quanta are ejected from the atom in specific directions, carrying all the energy. He mentions this idea briefly in a letter to Sommerfeld: "I am convinced that besides the directed energetic process, a kind of spherical wave is emitted, because of the possibility of interference for large-aperture angles. But . . . I am not convinced that what is being emitted immediately (the directed process) has an oscillatory character." He apparently had lengthy exchanges with Ehrenfest and Lorentz detailing his views, although he neither published nor spoke publicly of them. Lorentz himself included them (with credit to Einstein) in lectures at Caltech in 1922, and a letter from Lorentz to Einstein in November of 1921 survives, in which he recapitulates Einstein's proposal.

Basic idea: . . . Upon the emission of light there are two sorts of radiation. They are:

1. An interference radiation, which occurs according to the normal laws of optics but does not transmit any energy. . . . Consequently, they themselves cannot be observed; they just show the way for the energetic radiation. It is like a dead pattern that only comes to life through the energetic radiation.

2. The energetic radiation. It is composed of indivisible quanta of [energy] $h\upsilon$. Their path is given by the (vanishingly small) flow of energy from the interference radiation, and therefore they can never reach a spot where this flow is zero. . . . full interference radiation is formed [even if] . . . only a single quantum is emitted, which thus also can reach the receiving screen at only one spot. But this elementary instance is repeated countless times. . . . The various quanta now distribute themselves statistically . . . [so] that their average number at each point on the screen is proportional to the intensity of the incident interference radiation there. In this way the observed interference phenomenon is formed, consistent with the classical theory.

Einstein had realized that making sense of the behavior of light requires that that we describe radiation by an extended field or disturbance, which would *determine* the measurement of energy transfer at specific points in space, while at the same time the energy transfer would itself be composed of localized, and quantized, units. He came up with the picturesque name of "ghost fields" (*Gespensterfeld*) for his guiding disturbance, and it seemed that he was inclined to regard the particulate quanta as the "real" things while the extended, wavelike entity was relegated to a secondary, spectral existence. Nonetheless he took the idea quite seriously, and from 1918 on discussed it widely, so that not just Einstein's inner circle of confidants knew of it but also young students, such as Eugene Wigner,[1] who recalled that Einstein was "quite fond of it." Moreover, according to Lorentz, Einstein also foresaw that the ghost field would be used to determine probabilities for the light quanta. Continuing his description of Einstein's idea, he states: "It has to be assumed that at each reflection and diffraction, whenever an incident light beam is split into two or more beams, the *probability* that a light quantum takes one or the other path is proportional to the intensities of the motions of light along these various paths, *calculated according to the classical laws*" (italics added). "Classical laws" here refers to Maxwell's equations, which determine the reflection and diffraction of electromagnetic waves, and hence determine the probability that the quanta will go one way or the other. For individual photons, it seems, there *are* no deterministic laws of motion; only the properties of the guiding field are set by deterministic laws. These two ideas—that quantum particles are guided by an extended field, which allows them to "interfere with themselves," and that this extended field obeys definite laws but doesn't determine the fate of individual quanta—are key concepts in the modern quantum theory of radiation (and of matter). Usually they are attributed to Max Born,

[1] Wigner arrived in Berlin in 1921 at the age of nineteen and attended the famous physics colloquia of Berlin University. He witnessed firsthand the completion of quantum theory and would go on to win the Nobel Prize in 1963 for seminal applications of the theory, particularly for his identification and use of symmetry principles.

who applied them first to electrons, but he always credited Einstein as their source, even on the Nobel podium in Stockholm.

In 1917, at the same time as Einstein was beginning to develop his concept of "ghost fields" for quanta of light and trying not to give up the ghost himself from his various ailments, he somehow mustered the energy to look at the other side of the quantum dilemma: the quantum theory of atoms and electrons. Einstein had begun his work on quantum theory treating light and thermal radiation, but by 1907 had realized that Planck's law required that atoms and molecules must also obey a non-Newtonian mechanics, since their vibrational energy was necessarily quantized. At that time he chose to pursue a revised theory of radiation that would contain quanta of light, only to set the project aside in frustration in 1911. Instead it was Bohr who took the next step in atomic mechanics, in 1913, with his quantized electron orbits, followed closely by the elaborations of his theory due to Arnold Sommerfeld in 1915 that so impressed Einstein. Until the spring of 1917, in twelve years of work, Einstein had never proposed an equation to describe the quantum behavior of electrons, the key to understanding the atom. Rather remarkably, he took up exactly this challenge during his convalescence in March and April of 1917. His work would uncover a surprising deficit in the Bohr-Sommerfeld version of quantum theory.

The early version of the quantum theory of matter, which Bohr and Sommerfeld pioneered (and which is now known as the "old quantum theory"), had overcome the first major hurdle in generalizing the Planck-Einstein restriction that energy is quantized on the molecular scale. As we saw earlier, Planck and Einstein had imposed this restriction only on the simplest type of periodic motion, linear harmonic oscillation, whereas Bohr had generalized it through rather laborious arguments to the periodic motion of an electron "orbiting" the nucleus of the atom. Sommerfeld had realized that Bohr's prescription amounted to a more general rule than Planck's $\varepsilon = nh\nu$, but that the quantity actually quantized, both in electron orbits and in molecular vibrations, was the *action*.

Yes, *that* "action." Recall how Herr Planck always referred to his constant, h, as the "quantum of action." "Action" is a technical term in classical mechanics; it refers to a mathematical quantity that was introduced in the nineteenth century as a convenient and powerful way to think about any particle trajectory that obeys Newton's laws. The "action of a trajectory" is a number that physicists can calculate, but its meaning is not as intuitive and familiar as the concepts of mass, momentum, and energy introduced by Newton. Roughly it corresponds to the minimum amount of *time* that a particle can take in going from point A to point B, given the forces it experiences. Somewhat miraculously, the *actual* trajectory a Newtonian particle follows according to $F = ma$ manages to minimize this global measure, as if the particle had planned its path in advance. You can actually throw away Newton's laws and do all of classical mechanics just based on this "principle of least action."

And just like other quantities in classical mechanics, such as energy, action is also a continuously varying quantity, so there was no obvious way to break it up into smallest units. No way, that is, until the discovery of Planck's constant. Planck's constant is the first known fundamental physical constant that has the same units as action, the units of length times momentum. If I want to cut a rope into one-, two-, and three-foot sections, I need a ruler to measure this out; I could use this ruler to "quantize" the lengths of rope. Similarly, Sommerfeld had realized that Planck's constant was the natural "ruler" for quantizing periodic trajectories; the action along each trajectory had to be a whole number times h. You could throw out all the classically allowed electron orbits in the atom that didn't have action divisible by h with no remainder. For some reason, they are not allowed. This rule didn't explain *why* microscopic motion was quantized in a more fundamental sense than Planck's original hypothesis, but it allowed the quantum theory to describe a much wider range of motions than just simple harmonic oscillation. In particular Sommerfeld had been able to add elliptical electron orbits to Bohr's simple circular orbits, and even include the effect of relativity on the orbital energy as the electron whizzed by the nucleus. In fact this approach would occupy the community for the next seven years as more and more elaborate models were made for quantized orbits in the atom.

Einstein was intrigued enough by the elegance of Sommerfeld's method that immediately after finishing his new radiation theory, he turned his attention to it. But soon he would realize that there was a problem with this new theory of quantized actions, one that only he recognized, and which he set out to address while he was still barely off life support in early spring 1917.

The first clear indication of his new perspective comes in a letter to Besso on April 29, 1917: "yesterday I presented a little thing on the Sommerfeld-Epstein[2] formulation of quantum theory before the thinned ranks of our Phys. Society. I want to write it up in the next few days." By May 11 he was ready to present the finished product to the German Physical Society, a paper titled "On the Quantum Theorem of Sommerfeld and Epstein."

Einstein was motivated to write this paper by a core principle of his relativity theory: the laws of physics should not depend on the reference system of the observer or, in the case of the atom, on the choice of a coordinate system for describing electron orbits. He begins his paper by praising the Sommerfeld approach but then says "it still remains unsatisfying that one has to depend on [a specific choice of the coordinate system], because this probably has nothing to do with the quantum problem, *per se*."

Einstein was searching for a more general statement of the principle of quantizing action. And, armed with a much more advanced set of mathematical tools than he possessed before his development of general relativity theory, he found one: topological invariance. Topology is the branch of mathematics that studies the properties of objects that remain unchanged when the objects can be deformed continuously to another shape, but without breaking anything, so that the object retains the same number of "holes." In topology a doughnut and a coffee cup are considered to be the same shape, because the handle

---

[2] Paul Epstein was a young Jewish theoretical physicist from Poland who did his PhD with Sommerfeld and contributed to some further details of the Sommerfeld formulation. Einstein was rather impressed by him and helped him when he had difficulty finding a permanent job; he eventually immigrated to the United States with the help of Millikan and became a professor at Caltech.

of the cup corresponds to the hole in the doughnut and you could imagine taking the rest of the cup (if it were malleable) and squeezing it smoothly around the handle without ever breaking off a piece until it molded into a doughnut shape. In modern physics topological invariants have been found to play critical roles in many fundamental theories, but none had been found prior to Einstein's 1917 work. In a tour de force, Einstein, was able to show in a few short pages that the Bohr-Sommerfeld rule can be written in a manner that depends only on topological properties of the electron trajectories, essentially on how these orbits "wrap back" on themselves after each period of the motion. Since it is a topological property, it doesn't change if you rotate or deform the system, and hence it doesn't depend on any particular coordinate system.

So it looks like Einstein is satisfied that he has firmed up the Bohr-Sommerfeld theory with a little more rigorous mathematics and that is going to be the end of it. But there is more to the story. In a formulation that surely sounded as odd then as it does today, Einstein continues, "we now come to a very essential point *which I carefully avoided mentioning* during the . . . sketch of the basic idea." Actually, he points out, if the electron orbits are sufficiently complicated or irregular, so that there *is* no single period in which the trajectory wraps back on itself, then his method goes out the window, along with the whole approach of Bohr and Sommerfeld. Such complicated, irregular motion is now familiar to modern physicists, where it is described by the term *dynamical chaos*.

"One notices immediately," Einstein concludes, "that [motion of this type] excludes the quantum condition we have formulated." So there is a looming problem for the Bohr-Sommerfeld approach, which has baffled Einstein himself, the problem of quantizing chaotic motion.

Einstein's work was so advanced that few people understood it at the time; chaotic motion was not really appreciated until the advent of modern computers, and topological quantum rules were beyond everyone's ken. Despite the fame of its author, the problem identified in his article was ignored for more than fifty years, before a subfield of theoretical physics emerged known as "quantum chaos theory."[3] But the paper he had written was not unimportant at the time; in 1926, when Erwin Schrödinger wrote his epochal series of papers defining the wave equation of quantum mechanics, he included the following footnote: "The framing of the quantum conditions [in Einstein's 1917 paper] is the most akin, out of all the earlier attempts, to the present one."

[3] It was Einstein's 1917 paper that inspired the author to look again at Einstein's contributions to quantum theory, as described in the introduction (see A. Douglas Stone, "Einstein's Unknown Insight and the Problem of Quantizing Chaos," *Physics Today*, August 2005, pp. 37–43).

CHAPTER 23

FIFTEEN MILLION MINUTES OF FAME

But now I'm just about fed up to the teeth with relativity! Even such a thing pales when one is too occupied with it.
—ALBERT EINSTEIN TO ELSA EINSTEIN, JAN. 8, 1921

While Einstein struggled to regain his vitality and refocus his efforts on understanding the atom during the years 1917 and 1918, a singular confluence of social forces was occurring that would change his life irrevocably. On November 9, 1918, the German Reich surrendered, the kaiser abdicated, and Germany was thrown into political turmoil. The Great War would leave a residue of hatred, resentment, and disillusionment, which would affect international scientific cooperation for a decade. Yet within less than a year a British scientific expedition would catapult Einstein to a level of global fame unprecedented in the history of science. Arthur Stanley Eddington, the young Plumian Professor of Astronomy at Cambridge, led the expedition; he had been a conscientious objector during the war and was a person of undiplomatic stubbornness in whom Einstein would have recognized a kindred soul. When the Cambridge dons obtained for him a deferment from military service, on the grounds of his superior value as a scientist, he insisted on signing it with the stipulation that, had he been deemed of greater military value, he would have refused service anyway. Eddington had championed Einstein's theory of general relativity in England and, through the offices of the astronomer royal, Sir Frank Dyson, had been put in charge of the eclipse-observing expeditions that departed in February of 1919 to test Einstein's quantitative predictions of the bending of the path of starlight due to the gravitational field of the sun. With some eventfulness, in the end these expeditions confirmed Einstein's theory and reported their results at the historic meeting of the Royal Society on November 6, 1919.

The meeting, which the mathematician Alfred North Whitehead likened to "a Greek drama" in which the "laws of physics are the decrees of fate," placed Einstein on the same level as Sir Isaac Newton in the British scientific pantheon, leading to J. J. Thomson's famous pronouncement that Einstein's new framework was "one of the highest achievements of human thought."[1] Something about this new worldview, arcane but not encased in a protective layer of technical jargon, captured the general public's imagination as the "spectral lines" and "quanta" of atomic theory could never do. Space was curved; light was subject to gravity; all motion was relative. It was mind-bending stuff, but not incomprehensible techno-speak. One could conjure with it. And one did.

The *New York Times* captured the zeitgeist with one of its most whimsical headlines ever: "Lights All Askew in the Heavens: Men of Science More or Less Agog Over Results of Eclipse Expeditions. Einstein Theory Triumphs; Stars Not Where They Seemed or Were Calculated to be, but Nobody Need Worry. A Book for 12 Wise Men; No More in All the World Could Comprehend It."

By September 1920 relativity mania had swept the globe, leading Einstein to exclaim, "At present every coachman and every waiter argues about whether or not the relativity theory is correct." Einstein's first biographer described the ethos thus: "from the intellectual work of a quiet scholar a message of salvation had emerged. . . . No name was uttered as much during that time as that of this man. . . . Here was a man who had reached out to the stars, a man in whose theory one had to penetrate to forget one's earthly troubles." To have a transnational, apolitical figure to admire in the aftermath of the decade of

---

[1] This was often later misquoted as *the* highest achievement.

Ständer Nr. 50
78 Jahrgang
Berliner
Illuſtrirte Zeitung
Verlag Ullſtein & Co, Berlin SW 68
25 Pfg.

FIGURE 23.1. Cover of *Berliner Illustrirte Zeitung* of December 14, 1919, proclaiming Einstein as a new genius on the world stage.

senseless destruction satisfied a deep, if unanticipated, social need. And so Einstein became a symbol of man's better nature, a potential political force, an ethnic beacon to the Jews, an inspiration for revolutionary art, and a challenge to philosophers. His public lectures were sensational events; his opinions on all things were sought; he should write this or that commentary, give this or that speech, join this or that committee, for this or that worthy cause. All of which detracted from his true life's mission, to think deeply about nature and, more specifically, to finally get to the bottom of the conundrums of atomic theory. He wrote to his old friend Zangger in January 1921, "the fragmentation of one's intentions by the motley array of duties is crippling, especially for a person made more for concentration than for conformance."

While Einstein was soaring to unprecedented celebrity for a scientist, Germany was struggling to maintain a stable political order after its defeat in the Great War. The Weimar Republic, with its liberal constitution, was established in the summer of 1919 and for the next four years suffered repeated challenges from reactionary forces, led by the right-wing *Freikorps*. These forces fed the surge of anti-Semitism that followed the German surrender and signing of the Treaty of Versailles. Einstein, due to his newfound prominence, was a natural target to attack. A fringe group of scientists and engineers, led by an obscure engineer named Paul Weyland, attacked relativity theory as fallacious and undermining of the purity of German Science. Einstein, rather naively, thought that paying attention to this claque and refuting their claims with a combination of logical argument and sarcasm would be worthwhile. After a public meeting of the antirelativists in August of 1920, which Einstein himself attended, he attacked his critics in a newspaper article that he rapidly came to regret. "Everyone must, from time to time, make a sacrifice on the altar of stupidity . . . and I did so thoroughly with my article."

Indeed, in addition to drawing attention to this movement, which would otherwise have been insignificant, he also angered a respectable physicist, Philipp Lenard. Lenard was a Nobel laureate whose seminal experiments on the photoelectric effect had partly inspired Einstein's breakthrough work on quanta of light. Lenard had doubted the validity of relativity theory, writing an article critical of it in 1918, but at this point had never attacked Einstein personally. Lenard's stated problem with relativity theory was one that many could identify with: it violated "sound common sense." Lenard was not a theorist, and Einstein's colleagues would surely have realized that his criticisms were simply the result of an inability to grasp the theory's challenging abstractions. However, Weyland had appropriated Lenard's name for his group's use. and Einstein had assumed, perhaps with some justification, that Lenard shared their nonscientific motivations to attack him. Thus he wrote of Lenard by name: "[he] has so far achieved nothing in theoretical physics, and his objections to the general theory of relativity are of such superficiality that until now I had thought it unnecessary to answer them in detail."

Lenard naturally took offense at the insulting tone of Einstein's comments, prompting Sommerfeld to implore Einstein to make some conciliatory gesture, a request to which Einstein never acceded. In fact this first flurry of German attacks on Einstein's science, with its strong overtones of anti-Semitism, was never a major movement and was widely rejected by the German physics establishment of the time. Einstein himself felt at the time that his critics were no more than an annoyance, commenting famously, "I feel like a man lying in a good bed, but plagued by bedbugs." A decade later, however, the bugs would turn poisonous. Lenard would never forgive Einstein and henceforth opposed him at all turns. Eventually he would join the Nazi Party and lead its successful effort to oust Jewish scientists, including Einstein, from the Prussian academy, and eradicate "Jewish physics" from the textbooks.

Although in 1921, at the time of these early attacks, Einstein's scientific colleagues felt that it was unwise of him to engage with this rabble at all, they uniformly defended him. Most striking was a statement by von Laue, Nernst, and Rubens: "It cannot be our task to discuss in detail the unparalleled profound intellectual work which led Einstein to his theory of relativity. . . . What we do want to emphasize, and what was not touched upon in a single work yesterday [at the antirelativity meeting], is that *quite apart from Einstein's relativistic research, his other work already assures him of an immortal place in the history of science*" (italics added). So while the public either swooned or fumed over relativity theory, his Berlin colleagues had not lost sight of the fact that Einstein was the conceptual leader of the new atomic physics. And they had not abandoned the hope that he would yet come up with a true and complete quantum theory.

Einstein himself, however, seemed for the first time in his life easily distracted from his scientific research. Although he did not feel truly threatened by the anti-Semitic mood of the right wing in 1920, it did appear to rekindle a sense of ethnic identification with his Jewish brethren. In April 1920 he addressed a Jewish group as follows: "there is in me nothing which can be described as 'Jewish faith.' But I am happy to belong to the Jewish people." Not that he had any sympathy for religious observances, as he made clear to a rabbi with whom he had debated: "The [religious Jewish] community is an organization for the exercise of ritualistic forms that are remote from my opinions. I must take it for what it is today and not for what one might perhaps wish to see it transformed into. When I want to drive into town, I do not lay myself down in bed in the hope that it will grow wheels and become an automobile. . . . [However] I gladly vow . . . all kinds of efforts in the interest of individual Jews and Jewish communities."

In keeping with this pronouncement, less than a year later, despite his professed desire to focus on science, he immediately agreed to an invitation from Chaim Weizmann, the president of the World Zionist Organization, to accompany him to the United States to solicit funding for the planned Hebrew University in Jerusalem. Never mind that the trip's schedule would require his withdrawal from the first of the revived Solvay Congresses, on "atoms and electrons," and that among his many talents and interests fund-raising had never previously figured. To Haber (who greatly opposed his participation), he confessed: "I am not needed for my abilities, of course, but only for my name. Its promotional power is anticipated to bring considerable success thanks to our rich fellow clansmen of Dollaria (Einstein's nickname for the USA)." However, he regarded his participation as a moral duty: "Despite my declared international mentality, I do still always feel obliged to speak up for my persecuted and morally oppressed fellow clansmen, as far as it is within my powers. . . . The prospect of establishing a Jewish university delights me especially, after recently seeing from countless examples how perfidiously and unkindly fine young Jews are being treated here in the attempt to deprive them of educational opportunities." In fact, while Einstein created a sensation everywhere he went in America, adding to his legend, the trip was only modestly successful in its monetary goals.

This trip seemed to spark a wanderlust in Einstein that he had not previously demonstrated. During the next two years, in addition to the trip to the United States, mainly by his own choice, he would visit Holland, Austria, Czechoslovakia, England, France, Italy, Switzerland, Japan, Hong Kong, Singapore, Palestine, Spain, Sweden, and Denmark, not a program conducive to deep contemplation. However, in the summer of 1921, after returning from the United States, Einstein briefly turned his thoughts to the problem of light quanta once again.

While Einstein had by this time developed his notion of "ghost fields" to explain how particles of light could exhibit the interference effects associated with waves, he did not think that the net result for observations would be exactly as predicted by the classical theory of electromagnetic waves. Thus he sought an experimental test that would directly distinguish his theory, in which all energy was carried

FIGURE 23.2. Einstein and his second wife, Elsa Einstein, photographed on their trip to the USA in 1921. Library of Congress, courtesy AIP Emilio Segrè Visual Archives.

by individual light quanta, from the predictions of classical optics. In August of 1921 he thought he had found one, a "very interesting and quite simple experiment about the nature of light emission. . . . I hope I can carry it out soon."[2]

The idea was indeed quite simple, but flawed. Einstein assumed that light quanta would not show the Doppler effect when emitted from a moving atom (i.e., its frequency would not be shifted depending on the

---

[2] It is notable that Einstein, despite having become the paradigm of the pure theorist to the public, was still quite willing to contemplate actually setting up and performing such an experiment himself.

angle between its motion and the line of sight to the detector), whereas classical radiation was known to show such an effect. Thus he suggested imaging the light from moving atoms through a telescope lens with an inserted prismlike element, to differentially deflect the light of different frequencies. He calculated that the classical theory would give a deflected image and his quantum theory would not. He did not have to do the experiment himself, finding the seasoned professionals Walter Bothe and Hans Geiger quite happy to do these relatively easy measurements. They were completed in December of 1921; no deflection of the image was detected.

Einstein, again ecstatic, as he had been after his work on quanta in 1916, told Born the latest news. "Thanks to the excellent collaboration of Geiger and Bothe, the experiment on light emission is finished. Result: The emission of light by the moving [atom] is strictly monochromatic, whereas according to the undulatory theory, the color of the elementary emission ought to be different in different directions. Thus it is surely proven that the undulatory field has no real existence. . . . It is my most powerful scientific experience in years." However Einstein's exultation was short-lived. Independently, Ehrenfest and Max von Laue pointed out that Einstein had got the classical prediction wrong; both the classical and the quantum theory predicted no effect. Einstein redid the calculations himself and presented the amended conclusion a couple of months later: "in light of this theoretical result, deeper conclusions cannot be derived from the experiment concerning the nature of the emission process." He wrote again to Born, with characteristic self-deprecation, "I . . . committed a monumental blunder (experiment about the emission of light . . .). But one shouldn't take it too seriously. Death alone can save one from making blunders." He was clearly getting frustrated, once again, with the unyielding quantum: "I suppose it is a good thing that I have so much to distract me, else the quantum problem would have long got me into a lunatic asylum. . . . How miserable the theoretical physicist is in the face of nature."

The many distractions associated with his fame were compounded when, in June of 1922, right-wing extremists assassinated Walter Rathenau, the German foreign minister. Rathenau, the first Jew to hold that post, was a personal friend of Einstein's, and not only did his loss shake Einstein's equanimity, but in its aftermath multiple threats were received to his own life. He hid out in the country, sheltered by a rich friend, Hermann Anschütz, and even toyed briefly with quitting physics research and working as an engineer, with "a downright normal human existence" and the "welcome chance of practical work." This fanciful notion vanished in a few days, but he avoided the German physics meeting in Leipzig, bitterly disappointing the twenty-one-year-old Werner Heisenberg, who attended with the hope of meeting him.[3] In a letter to Solovine, his longtime friend and "Olympia Academy" alumnus, Einstein recounted: "I am constantly being warned, I have given up my lecture course and am officially absent although I am really here. Anti-semitism is very strong." He felt it was fortunate that he had "the opportunity of a prolonged absence from Germany," as he had committed himself to an extended lecturing trip to Japan and points east in October of 1922.

However, an unexpected wrinkle developed in his plans when he was informed by Svante Arrhenius, who still ruled the Physics Committee for the Nobel Prize, that "it will probably be very desirable for you to come to Stockholm in December, and if you are in Japan that will be impossible." The Nobel committee had finally been moved to award the most famous scientist since Newton its stamp of approval. Characteristically, Einstein was not disposed to dance to the tune of the establishment, particularly for an institution that had passed him over for so long; despite Arrhenius's hints that Einstein's absence might put the final vote in doubt, he replied that he was "quite unable to postpone the journey." So off he went to Japan by ship, on schedule, receiving news of his award somewhere en route on an unknown date, as he failed to even note the event with an entry in his travel diary. Nonetheless, the citation on the award surprised many:

---

[3] Sommerfeld, with whom Heisenberg was studying, had promised to introduce him to Einstein, who had been Heisenberg's idol from his high school days. Heisenberg was appalled to find an anti-Semitic leaflet attacking Einstein thrust into his hand as he entered the hall where von Laue was delivering the lecture on Einstein's behalf.

"for his services to theoretical physics and especially for his discovery of the law of the photoelectric effect." The Nobel committee had found relativity theory too uncertain and controversial for recognition but instead came to rest on the one work by Einstein that he himself considered "revolutionary."

While there was much irony in this development, in fact the citation was carefully crafted to recognize the *empirical* law of the photoelectric effect, and not the underlying theory. This law had already been confirmed in great detail by the American physicist Robert Millikan by 1916, and by many other experiments subsequently, so it could no longer be doubted. Millikan himself wrung his hands at his own results, saying "I spent ten years of my life testing that 1905 equation of Einstein and, contrary to all my expectations, I was compelled to assert its unambiguous experimental verification in spite of its unreasonableness, since it seemed to violate everything we knew about the interference of light." The Nobel citation said nothing about quanta of light, a concept that was still rejected by the overwhelming majority of physicists. In fact Bohr, despite his deep admiration for Einstein, could not resist a rather biting joke at Einstein's expense, saying that if he received a telegram from Einstein confirming the existence of light quanta, he would point out that the telegram itself, transmitted by electromagnetic waves, was proof against them.

Einstein did not return to Germany until the spring of 1923, and he went to collect his Nobel Prize in the summer. By this time he was involved in a different scientific quest, his famous attempt to unify the gravitational and electromagnetic forces by means of a "unified field theory." However, he was very hopeful that such a theory would itself point the way to the solution of the quantum problem, even submitting a paper along those lines to the Prussian Academy in December of that year. In 1924, in a radio address, he informed the public, who only wanted to hear about relativity theory, that he was really focused on something else:

The other great problem that I have been concerned with since about 1900 is that of radiation and the quantum theory. Stimulated by the work of

Wien and Planck, I recognized that mechanics and electrodynamics were in irresolvable contradiction with experimental facts, and so I have helped in creating the complex of ideas known by the name of quantum theory, which has been developed so fruitfully particularly by Bohr. I shall probably devote the rest of my life to the fundamental clarification of this problem, however slight the prospects are for attaining the goal may be.

In fact the goal was now rather close, although it would not take the form Einstein hoped, and he was about to make his final historic contribution to reaching it.

CHAPTER 24

THE INDIAN COMET

Respected Sir:

I have ventured to send you the accompanying article for your perusal and opinion. I am anxious to know what you think of it. You will see that I have tried to deduce the coefficient $8\pi v^2/c^3$ in Planck's Law independent of the classical electrodynamics, only assuming that the ultimate elementary region in the phase-space has the content h^3.

This letter to Einstein from an unknown Indian scientist, received in early June, 1924, initiated one of the most extraordinary episodes in the modern history of science, culminating in Einstein's final historic contribution to the structure of the new quantum theory. At the time of his writing, Satyendra Nath Bose was a thirty-year-old Reader (roughly equivalent to the rank of associate professor) at Dacca University in East Bengal. His previous five research papers had made no impact at all on contemporary research, and he had recently been informed that, due to a funding cutoff to the university, his appointment would not be extended more than a year. Moreover, the paper he was sending to Einstein had already been submitted for publication, and rejected, by the English journal *Philosophical Magazine*. He had admired Einstein for many years and had even produced a rather undistinguished translation of Einstein's papers on general relativity into English, for distribution in India. Thus, through some combination of veneration and chutzpah, he hit upon the idea of sending this paper, which related closely to Einstein's 1916 work on radiation theory, directly to the master, with an astonishing request:

I do not know sufficient German to translate the paper. If you think the paper worth publication, I shall be grateful if you arrange for its publication in Zeitschrift fur Physik. Though a complete stranger to you, I do not feel any hesitation in making such a request. Because we are all your pupils though profiting only by your teachings through your writings.

Yours faithfully, S. N. Bose.

Einstein by that time, as we have seen, was not just the most famous scientist of his time; he was one of the best-known individuals on the entire planet. He was deluged by letters from strangers, wanting his opinion on everything under the sun, while at the same time struggling to keep up his voluminous scientific correspondence with the large community of physicists with whom he had personal and professional relations. In addition, Einstein spoke very little English and had not been able to deliver his prestigious lectures in England during his visit in 1921 in the native tongue of his audience.[1] The a priori probability that S. N. Bose's paper would end up in the circular file, his work and his name lost to posterity, was extremely high.

But that is not what happened. Einstein read the paper shortly after its arrival, translated it, and sent it to the German journal *Zeitschrift für Physik* on July 2, 1924, with his strong endorsement. There it was subsequently published, with a note from Einstein appended. "In my opinion Bose's derivation of the Planck formula signifies an important advance. The method used also yields the quantum theory of the ideal gas, as I will work out in detail elsewhere." In fact, shortly thereafter Einstein translated a *second* paper, which he had received from Bose on the heels of the first one and which he sent on to the journal by July 7, 1924. This one did not elicit such a favorable opinion from the great man, and he published a critical comment along with it, while nonetheless supporting its publication.

---

[1] Even after he moved to the United States in 1933, Einstein never fully mastered the language, and one of his close collaborators, Leopold Infeld, said he functioned with "about 300 words, which he pronounced very weirdly."

With these magnanimous gestures from the sage, the die was cast. Bose would go on to become one of the most famous names in the history of modern physics. The term "boson" is used for one of the two fundamental categories of elementary particles in modern physics[2] because such particles obey novel statistical laws first employed, although not announced, in Bose's initial paper sent to Einstein. This category includes Einstein's photons (light quanta) as well as roughly half of the atoms in the periodic table. But Bose's discovery, like that of Planck twenty-four years earlier, was not as clear-cut as it has been portrayed, and again it would take Einstein to find the radical implications in it.

S. N. Bose was born into the rising middle class of English-educated Indians in 1894 in Calcutta. He father, an accountant, had a wide range of intellectual interests that he transmitted to his son, who, in addition to his great aptitude for mathematics, became deeply interested in poetry, music, and diverse languages. When he matriculated at the Presidency College in Calcutta in 1909, however, he chose to study science, at least partly because of its potential utility to the future Indian nation, as a wave of nationalism swept through his generation. His cohort was "a particularly brilliant lot—the famous 1909 batch of Presidency College . . . [which] in all its history has not seen the likes . . . since." Bose completed his BSc in 1913 and MSc in 1915, taking first place in both examinations; but there was no obvious avenue for obtaining a doctorate, and the professorial ranks were still reserved for third-rate English academics at that time. Bose therefore went through a period as a striving outsider, not dissimilar to the career of Einstein at the same age.

He married early (prior to graduation) but, contrary to custom, refused to accept a dowry or other financial support from his wife's family. Already responsible for a wife and son shortly after he received his MSc, he spent a year eking out a living through private tutoring, while trying to work toward a PhD in mathematics with a well-known

---

[2] Bosons are the force-carrying particles of the fundamental fields. The most recent confirmed member of this group is the Higgs boson, related to the electroweak interaction. Atoms are not fundamental particles, but are composites of quarks and electrons that can still behave statistically as bosons.

professor, Ganesh Prasad. Prasad was noted for his aggressive criticism of prospective students and of their previous teachers, which typically cowed the candidates into silence. But Bose was "notorious for plain speaking." In an echo of Einstein's conflicts with authority figures such as Weber, Bose "dared to counter his adverse criticisms" and was summarily dismissed from consideration for PhD work with the comment, "you may have done well in the examination, but that does not mean you are cut [out] for research." "Disappointed, I came away [and] decided to work on my own," Bose recalled.

Like Einstein, he was turned down for low-level teaching jobs before being offered an entry-level lectureship in the new University College of Science in Calcutta, whose founder, Sir Asutosh Mookerjee, began hiring the cream of young Indian scientists, including the future Nobel laureate C. V. Raman. It was at the College of Science that Bose first began to learn about the exciting developments in physics in Europe associated with the names of Planck, Einstein, and Bohr. Here also he and his close friend, the physicist Meghnad Saha, obtained and translated important German works of physics, including Einstein's papers on relativity theory.

In 1921 Bose accepted a faculty position at the new University of Dacca, in East Bengal, recently established by an ambitious vice-chancellor, P. J. Hartog. At Dacca he "spent many sleepless nights" trying to understand the Planck law, while at the same time teaching it to his students. He felt an obligation to present something to them that he himself found clear and consistent: "As a teacher who had to make these things clear to his students I was aware of the conflicts involved. . . . I wanted to know how to grapple with the difficulty in my own way. . . . *I wanted to know.*" In late 1923 he hit upon his new approach to deriving the law and sent off a manuscript to the *Philosophical Magazine*, where he had previously published papers on quantum theory, only to receive the reply, in the spring of 1924, that the referee's decision had been negative. It was at this point that Bose took the bold step of sending the paper to Einstein, a strategy so speculative that its success appeared to violate the very principle of maximum entropy employed in the paper itself!

It was remarkable, but nonetheless true, that Planck's blackbody formula remained somewhat mysterious a full twenty-four years after Planck's initial derivation. It was not that anyone doubted any longer the validity of the formula, but the tortured reasoning Planck had used to derive it left physicists unsatisfied for decades. That is why Bose's paper, titled "Planck's Law and the Quantum Hypothesis." was of interest to Einstein and others. As we saw earlier, Planck had been reluctant to treat radiation directly with statistical mechanics, and instead, using classical reasoning, he related the mean energy of radiation at a given frequency to the mean energy of idealized vibrating molecules ("resonators"). He then calculated the entropy of these resonators by introducing into the counting of states his quantized energy "trick." A key factor in totaling up the number of allowed states in Planck's method, which was not appreciated for quite some time, was that one could treat the units of energy belonging to each resonator as indistinguishable quantities (i.e., if resonator one had seven units of energy $h\upsilon$ and resonator two had nine units, one didn't have to ask *which* units they were). In 1912 Peter Debye, an outstanding young theorist and future Nobel laureate, rederived the Planck law, not by counting resonator states, but by counting states of classical electromagnetic waves that could fit in the blackbody cavity and then ascribing to them the same average energy that Planck had assigned to each resonator. The counting of the number of allowable *waves* in the cavity led to the factor in the Planck radiation formula, $8\pi\upsilon^2/c^3$, to which Bose alludes in his letter to Einstein. This factor was very easy to find from classical wave physics but very hard to find from quantum principles, hence Bose's emphasis on having found a quantum route to it.

It was clear that Einstein had been troubled by Planck's derivation from his earliest works, but not until 1916 had he even tried to justify this law, succeeding marvelously with his "perfectly quantic" paper, which introduced the concepts of spontaneous and stimulated emission of photons, as well as providing strong arguments for the reality of photons. One major reason that Einstein was so happy with this work, and even called it "*the* derivation" of Planck's formula, was that it did *not* at any point use the strange counting of the distribution of energy units that Planck had employed. Instead he managed to get the same answer by a different route, based on Bohr's quantized atomic energy levels and his own plausible hypotheses about the balancing out of all emission and absorption processes. Bose, though, was not completely happy even with this method and claimed to have found an even more ideologically pure derivation.

Bose's paper is concise in the extreme, running to less than two journal pages. He begins the work by laying out his motivation for presenting yet another approach to the Planck law. "Planck's formula . . . forms the starting point for the quantum theory . . . [which] has yielded rich harvests in all fields of physics . . . since its publication in 1901 many types of derivations of this law have been suggested. It is acknowledged that the fundamental assumptions of the quantum theory are inconsistent with the laws of classical electrodynamics." However, Bose continues, the factor $8\pi v^2/c^3$ "could be deduced only from the classical theory. This is the unsatisfactory point in all derivations." Even the "remarkably elegant derivation . . . given by Einstein" ultimately relies on some concepts from the classical theory, which he identifies as "Wien's displacement law" and "Bohr's correspondence principle," so that "in all cases it appears to me that the derivations have insufficient logical foundation."

Einstein did not agree with this criticism and even took time out during his first, quite friendly letter to Bose to dispute it: "However I do not find your objection to my paper correct. Wien's displacement law does not presuppose [classical] wave theory, and Bohr's correspondence principle is not used. But this is unimportant. You have derived the first factor [$8\pi v^2/c^3$] quantum-mechanically. . . . It is a beautiful step forward." On both points Einstein was correct.[3] Bose had come up with a more direct method of getting the result, the first to use only the photon concept itself, a tremendously appealing simplification.

---

[3] The displacement law, which constrains the form of the Planck law but does not determine it, follows from general principles of thermodynamics and doesn't require Maxwell's equations. And Einstein had not used Bohr's correspondence principle, which at that time related only to the mechanics of atoms, but had simply used the known coefficient of the Rayleigh-Jeans law. However, the latter did require some form of counting of waves, so at least the thrust of this objection by Bose had some merit.

Ever since Einstein's 1905 paper on light quanta, there had been a glaring logical problem with taking quanta seriously as elementary particles. In dealing with the statistical mechanics of a gas of molecules, it is possible to derive all the important thermodynamics relations, such as the "ideal gas law" ($PV = RT$),[4] without ever specifying any other system with which the gas molecules interact. It is enough to simply say that there exist other large systems ("reservoirs") with which the gas can exchange energy. Then counting the states of the gas, using the classical method (no h!) pioneered by Boltzmann, leads to both the entropy and the energy distribution of the gas molecules, and eventually to all the known relations.[5] The very same approach appeared to fail for quanta of light; it led to Wien's incorrect radiation law, and not Planck's. This was a major problem, which, along with the difficulty in explaining the interference properties of light, led to the consensus that light quanta weren't "real" particles but some sort of heuristic construct. This consensus had survived even the awarding of the Nobel Prize to Einstein for the photoelectric effect. Bose's work shows how to escape from the first of these dilemmas.

Bose sets out to count the possible states, W, into which many light quanta of energy $h\upsilon$ and momentum $h\upsilon/c$ can be distributed according to quantum principles. From this, by a variant of Planck's method, he obtains the average entropy and energy of the photon gas.[6] Step one is to consider a *single* light quantum, with energy $h\upsilon$ and momentum $h\upsilon/c$. If a photon were to be treated as a real, classical particle, one ought to be able to specify its state at each time by its position and its momentum. Physicists refer to such quantities as *vectors*, since they

---

[4] Here P is pressure, V is volume, T is temperature, and R is the gas constant, related to Boltzmann's constant, k, discussed earlier. A special case of this law is Boyle's law, that gas pressure is inversely proportional to its volume at fixed temperature.

[5] In the next chapter we will see that there actually were subtle flaws in this method, which Einstein would discover and then, through the application of Bose's ideas, show that the real ideal gas will deviate from the classical behavior found by Boltzmann. But these deviations were not yet detectable, and the problem was not with the concept of a gas connected to unspecified reservoirs but with Boltzmann's counting method.

[6] Henceforth I will use "photon" (the modern term) and "light quanta" (Einstein and Bose's term) interchangeably. The photon gas is the standard modern terminology.

carry both a magnitude and a direction (e.g., the photon is 5 blocks northeast, with its momentum [always parallel to its direction of motion] due south). The momentum for a massive particle is just its mass times its velocity vector (when its speed is much less than c); but a photon's speed (magnitude of velocity) is always equal to c, and Einstein has shown (e.g., in his 1916 work) that the magnitude of its momentum is $h\upsilon/c$ (not m times c, since the photon has zero mass).

Counting the position states of the photon is not the hard part of Bose's argument; one assumes that the photon gas is enclosed in a box of volume V and it can be anywhere within V, with equal probability (this point was used by Einstein in his original arguments for the photon concept when analyzing the blackbody entropy back in 1905). Therefore Bose focuses on counting *momentum* states. Since the possible directions of photon motion are continuous and hence infinite, he has to employ an idea already proposed by Planck as early as 1906. Planck's constant, h, defines a quantum limit on the smallest difference in momentum that can be resolved.[7] Since all photons of frequency υ have the same magnitude of momentum, $h\upsilon/c$, which is assumed equally likely to point in any direction, Bose can count their states by tiling the surface of a sphere of radius $h\upsilon/c$ with these "Planck cells." From basic geometry and the assumption that the spherical shell is only one cell thick, he is then able to find the number of states,[8] $8\pi\upsilon^2/c^3$.

Up to this point, what Bose has done is logically appealing but not historic. It was his next step that caused his paper to become "the fourth and last of the revolutionary papers of the old quantum theory." He still has to obtain the Planck form for the radiation law and not the

---

[7] Actually the relevant unit of "resolution" is a cell simultaneously in position and momentum, known as a "phase space volume" (mentioned in Bose's first letter to Einstein); this cell has volume h^3 (again as mentioned in Bose's letter).

[8] In this argument he includes at the end a final factor of 2 he needs to recover the correct coefficient by assuming that the concept of polarization of EM waves can be extended to photons. Since polarization is a property of waves and not particles, this step was not completely rigorous, as Einstein pointed out to Bose. Much later Bose would claim that he had proposed that the photon has a spin with two possible states, now the accepted theory, but that Einstein had rejected this view and "crossed it out" of the first paper.

Wien form. For this next, crucial step he has to calculate how many physically distinct ways there are to put *many* photons at the same time into these available states. But Bose does not appear to realize that the next step is the big one; instead he seems to think that the previous one was the most significant. He begins the relevant paragraph by saying, "It is now very simple to calculate the thermodynamic probability of a macroscopically defined state." After a few definitions, he unveils his answer, a rather obscure combinatorial formula bristling with factorial symbols. This is the key intermediate step. From here on, finding the Planck formula is inevitable and just involves straightforward manipulations, within the competence of scores of his contemporaries.

This paltry written record leaves an enormous historical question. To what extent did Bose understand the key concept in his "revolutionary" derivation? For buried in Bose's factorial formula is a very deep and bold assertion. This formula implies that interchange of two photons in the photon gas does *not* lead to different physical states, unlike the standard, classical, Boltzmann assumption for the atomic gas. Boltzmann, and everyone else after him, assumed that even though atoms were very small and presumably all "looked" the same, one could imagine labeling them and keeping track of them. And if photons were particles like atoms, one should be able to do the same thing. Photon 1 having momentum toward the north and photon 2 having momentum toward the south was a physically different condition from photon 2 north, photon 1 south. Bose, without saying a word about it in his paper, implicitly denied this was so!

Late in his life Bose was asked about this critical hypothesis concerning the microscopic world, which followed from his work. He replied with remarkable candor:

I had no idea that what I had done was really novel. . . . I was not a statistician to the extent of really knowing that I was doing something which was really different from what Boltzmann would have done, from Boltzmann statistics. Instead of thinking about the light quantum just as a particle, I talked about these states. Somehow this was the same question that Einstein asked when I met him: How had I arrived at this method of deriving

Planck's formula? Well, I recognized the contradictions in the attempts of Planck and Einstein, and applied statistics in my own way, *but I did not think that it was different from Boltzmann statistics* [italics added].

Recall that in his derivation Bose was guided by the knowledge of the end point, the precisely known formula for the Planck radiation law. So he did not have to convince himself in advance that his counting method was justified; it turned out to be the method that gave the "correct" answer, so it would seem to be justified a posteriori. He appears to have missed the fact that in asserting this new counting method, he had made a profound discovery about the atomic world, that elementary particles are indistinguishable in a new and fundamental sense.

Einstein, despite his initial enthusiasm for the prefactor derivation, very quickly grasped that the truly significant but puzzling thing about the Bose work was the postclassical counting method. Soon after receiving Bose's paper, he expressed this in a letter to Ehrenfest: "[the Bose] derivation is elegant, but the essence remains obscure." He would pursue and ultimately elucidate this obscure essence for the next seven months.

Bose, not having fully appreciated the novelty of his first paper, placed a great deal of emphasis on his second paper, which was dated June 14, 1924, and sent to Einstein immediately on the heels of the first one. This paper is not a rederivation of a known result, as is his previous paper, but an attempt to reformulate the quantum theory of radiation, in direct contradiction to Einstein's classic 1916 work. The paper, titled "Thermal Equilibrium of the Radiation Field in the Presence of Matter," proposes a bold hypothesis. While there still is a balance between quantum emission and absorption leading to equilibrium, the emission process is assumed to be *completely* spontaneous and independent of the presence or absence of external radiation. Bose has eliminated Einstein's hypothesis of stimulated emission, which he refers to as "negative irradiation," saying it is "not necessary" in his theory. To make things balance out, he then has to assume that the probability of absorption also has a different and more complicated dependence on the energy density of radiation than Einstein assumed.

Einstein, who was a virtuoso at finding absurd implications of flawed theories since his days at the patent office, published a decisive critical note appended to his translation of this paper. First, he notes that Bose's hypothesis "contradicts the generally and rightly accepted principle that the classical theory represents a limiting case of the quantum theory. . . . in the classical theory a radiation field may transfer to a resonator positive or negative energy with equal likelihood." Second, Bose's strange hypothesis about the nature of absorption implies that a "cold body should absorb [infrared radiation] less readily than the less intense radiation [at higher frequencies]. . . . It is quite certain that this effect would have been already discovered for the infrared radiation of hot sources if it were really true." Because of these compelling arguments, Bose's only published attempt to *extend* quantum theory had no influence on the field and is solely of historical interest.

Nonetheless, Einstein's recognition of Bose's first paper transformed his professional situation in an instant. Einstein's supportive postcard to Bose congratulating him on his "beautiful step forward" was shown to the vice-chancellor at Dacca, and it "solved all problems." Bose recalled, "that little thing [the postcard] gave me a sort of passport for [a two-year] study leave [in Europe] . . . on rather generous terms. . . . Then I also got a visa from the German consulate just by showing them Einstein's card."

By mid-October of 1924 Bose had arrived in Paris and was introduced to the noted physicist Paul Langevin, who was a personal friend of Einstein's. Bose immediately wrote to Einstein, asking Einstein's opinion on his second paper (which he was unaware had already been published with Einstein's assistance) and expressing his desire to "work under you, for it will mean the realization of a long cherished hope." Einstein quickly replied, "I am glad I shall have the opportunity soon of making your personal acquaintance," then summarized his reasons for rejecting Bose's conclusions in the second paper and concluded by saying, "we may discuss this together in detail when you come here."

Despite the warmth of Einstein's reply, Bose was reluctant to move on to Berlin immediately, in part because Einstein's potent critique had made him unsure of his new proposal, which he wanted to refine further. In addition he seems to have found the change of cultures challenging; he decided to settle in Paris in the company of a local circle of Indian compatriot intellectuals. He justified this as follows: "because I was a teacher . . . and had to teach both theoretical and experimental physics . . . my motivation then became to learn all about the techniques I could in Paris . . . radioactivity from Madame Curie and also something of x-ray spectroscopy." An interviewer of Bose in 1972 noted that "even more than forty years later one still has the impression that Bose was terribly intimidated by most Europeans." This no doubt contributed to his disastrous interview with Madame Curie, concerning joining her lab. She had hosted a previous Indian visitor and had fixed it in her mind that the collaboration had failed because of his poor French. Thus she conducted her first interview with Bose entirely in English, and while welcoming him warmly, firmly insisted he would need four months of language preparation before starting work. Although "she was very nice," Bose, who had studied French already for ten years, found no opportunity to interrupt her monologue. And so "I wasn't able to tell her," he later explained, "that I knew sufficient French and could manage to work in her laboratory."

Despite this missed opportunity, Bose tarried in Paris nearly a full year learning x-ray techniques before working up the courage to move on to Berlin in October of 1925 and finally meeting Einstein a few weeks later. In the intervening year, Einstein had taken up Bose's novel counting method and extended it to treat the quantum ideal gas, leading to truly remarkable discoveries, about which Bose was unaware. For Bose, "the meeting was most interesting. . . . he challenged me. He wanted to find out whether my hypothesis, this particular kind of statistics, did really mean something novel about the interaction of quanta, and whether I could work out the details of this business." During Bose's visit, Werner Heisenberg's first paper came out on the new approach to quantum theory known as matrix mechanics (of which we will hear more later). Einstein specifically suggested that Bose try to understand "what the statistics of light quanta and the transition probabilities for radiation would look like in the new theory."

However, Bose was not able to make progress. He seems to have had a difficult time assimilating these rapid new developments and wrote somewhat despairingly to a friend: "I have made an honest resolution of working hard during these months, but it is so hard to begin, when once you have given up the habit." Bose received extensive access to the scientific elite of Berlin through Einstein's patronage and experienced the whirlwind of excitement around the revolution in atomic theory. But no publication resulted from his stay in Europe, and in late summer of 1926 he returned to Dacca. By then the new quantum mechanics had passed him by.

Bose became a revered teacher and administrator in his subsequent career in India, but he published little, and nothing that has

FIGURE 24.1. S. N. Bose photographed in Paris in 1924. Courtesy of Falguni Sarkar, SN Bose Project, www.snbose.org.

survived in the scientific canon. He continued to write to Einstein, periodically, and late in Einstein's life tried to visit him in Princeton, but he was denied a visa because "your senator McCarthy objected to the fact that I had seen Russia first." He eulogized Einstein eloquently upon his death: "His indomitable will never bowed down to tyranny, and his love of man often induced him to speak unpalatable truths which were sometimes misunderstood. His name would remain indissolubly linked up with all the daring achievements of physical sciences of this era, and the story of his life a dazzling example of what can be achieved by pure thought." For his own part, Bose seemed content with his role in scientific history, summing up his career aptly: "On my return to India I wrote some papers . . . they were not so important. I was not really *in* science any more. I was like a comet, a comet which came once and never returned again."

CHAPTER 25

QUANTUM DICE

Just under two years before Einstein's famous rejection of the new quantum mechanics with the memorable phrase "I . . . am convinced that [God] is not playing at dice," Einstein himself, inspired by Bose, changed the laws governing the playing of dice. Bose had unwittingly introduced a new method of counting the states of a physical system in order to derive the Planck law from direct consideration of a gas of light quanta, treated as particles, not waves. It was Einstein who would now explain and extend this new representation of the microscopic world to resolve long-standing paradoxes in gas theory and to reveal dramatic and previously undreamed-of behavior of atomic gases at low temperature.

Einstein had become renowned as the young genius of statistical physics ("Boltzmann reborn") through the sponsorship of Nernst fifteen years earlier, when Nernst realized that only Einstein's radical quantum theory of the specific heat of solids would validate his own famous "heat theorem": that the entropy of all systems should tend to zero as the temperature goes to zero. This fortunate confluence of Einstein's quantum principles and the interests of the most powerful scientist in Germany had played a significant role in winning Einstein his comfortable Berlin existence, free of teaching and administrative responsibilities. Einstein was now to add a sequel to this story.

Nernst had been arguing since 1912 that something similar to Einstein's freezing of particles into their lowest quantum states must occur for a gas of atoms or molecules at sufficiently low temperatures. However, *how* this would come about for a gas was a major puzzle. Gas particles are free to move over macroscopic distances, unlike electrons bound to atomic nuclei. In quantum theory, the larger the volume over which a particle is constrained to move, the smaller is its lowest allowed energy level, known as its "ground state." When you worked out this amount of energy for a gas particle in a container of human scale, it was absurdly small compared with the thermal energy scale, kT, *even* when the temperature was reduced to a few degrees above absolute zero.[1] So gas particles did not freeze out in the same way that vibrations of a solid did, according to the now-accepted form of Einstein's 1907 theory, refined by Peter Debye. Planck, Sommerfeld, and others had analyzed gases from the point of view of quantum mechanics and had failed to find an entropy function that obeyed Nernst's theorem. Of course, as we have learned, entropy is all about counting possibilities, and all the previous attempts had counted possibilities from the same point of view as Boltzmann. This point of view regarded atoms or molecules, even if identical in appearance, as distinct, distinguishable entities, in the same self-evident sense that a well-made pair of dice are identical in appearance but are distinct entities. It was this very obvious but very fundamental extrapolation from our macroscopic world that Bose had implicitly denied, and which Einstein would now explicitly deny. Einstein would yet again tell the world that our collective intuition about commonsense properties of the natural world is mistaken.

Einstein must have realized immediately, upon reading it, that Bose's approach would allow him to resolve the decade-old problem of the quantum ideal gas. For on July 10, 1924, just a few weeks after receiving, translating, and submitting Bose's first paper for publication, Einstein was reading his own paper, titled "Quantum Theory of the Monatomic Ideal Gas," to the Prussian Academy. This was the work to which he had alluded already in his famous "Comment of the Translator" published at the end of Bose's first paper: "The method used here also yields the quantum theory of the ideal gas, as I will

---

[1] Such extremely low temperatures were now available to physicists since 1908, following the Nobel Prize–winning research of the Dutch physicist H. Kammerling Onnes on liquefying and cooling helium. Einstein and Onnes were well acquainted, having met at the First Solvay Congress, and interacted frequently during Einstein's regular visits to Leiden.

show in another place." He minces no words in his opening to the gas theory paper: "A quantum theory of the . . . ideal gas free of arbitrary assumptions did not exist before now. This defect will be filled here on the basis of a new analysis developed by Bose. . . . What follows can be characterized as a striking impact of Bose's method."

In the next section of the paper Einstein directly follows the same computational method that Bose applied to the gas of light quanta, now applied to a gas of atoms. The analysis differs in only two significant ways. First, as Bose correctly assumed, a gas of photons loses energy as it is cooled simply by the disappearance of photons. As we already know, according to quantum theory, a photon is absorbed and disappears when it excites an electron in an atom to a higher energy level (and similarly can appear out of nothing when that atoms reemits energy and the electron quantum jumps back down to the lower level). This is the process that Einstein analyzed in detail in his famous 1916 paper, which set Bose on his quest for the perfect derivation of Planck's law. The total number of photons inside a box decreases as the box is cooled. The situation for an atomic gas is quite different. Atoms cannot just disappear,[2] so in analyzing the atomic gas, unlike the photon gas, Einstein has to add the constraint that the number of gas particles is fixed. Second, unlike photons, which always move at the speed of light, gas particles can lose energy simply by slowing down. For an *ideal* gas, which is the case Einstein is considering, in fact *all* the atomic energy is in the kinetic energy of motion of the atoms.[3]

With the constraint of a fixed number of atoms, Einstein correctly derives all the fundamental equations of the quantum ideal gas, which turn out to be substantially more complicated than those for the photon gas and do not lead to a relatively simple formula ("equation of state") analogous to $PV = RT$, which describes the classical gas. Thus

---

[2] At very high energies there are processes in which massive particles, such as an electron and a positron, annihilate one another and disappear, but such processes are not relevant here.

[3] It is a good approximation to neglect the interaction energy between atoms, based on the assumption that the gas is dilute enough that most of the time atoms are well separated, so that their mutual interaction is weak.

Einstein has to employ a subtler mode of analysis of these equations. He identifies a "degeneracy parameter," a ratio of variables that, if much larger than one, will lead back to the classical equation, $PV = RT$, but, if it approaches one, will lead to a new and different gas law. Thus this parameter measures the "quantumness" of the gas, and since it decreases with decreasing temperature, the theory implies that quantum effects will become more and more important the colder the gas becomes. To see if these deviations from the usual law will be observable, he plugs in numbers and finds that for a typical gas at room temperature this degeneracy parameter is very large, about 60,000, consistent with observations that all gases at room temperature obey the classical law ($PV = RT$) extremely well, and that the gas molecules obey the equipartition theorem, $E_{mol} = 3kT/2$, with no hint of quantum effects.

Next he analyzes what form the quantum corrections to the usual behavior will take if the temperature, and hence the degeneracy parameter, can be decreased to the point where deviations from classical behavior are no longer too small to observe. Sure enough, he finds that the energy per particle begins to drop below the equipartition value; so some precursor of quantum freezing *is* beginning to take place in the gas despite its macroscopic scale. Thus his results hint that Bose's statistical method will restore Nernst's theorem even for the ideal gas.

It seems unlikely that Einstein realized the full implications of Bose's approach when he wrote this first paper, since when he introduces Bose's new counting method, just like Bose, he does not explain or defend it with even a single sentence. Evidently the realization of just how strange the implications of this new statistical theory are had not yet fully dawned on Einstein. Hence his comment to Ehrenfest, in a letter sent two days *after* presenting his paper to the academy, admitting that the essence of the new approach is "still obscure." By September, two months later, he hints in a letter to Ehrenfest that things are becoming clear but that the implications are so strange as to raise doubts: "the theory is pretty, but is there also any truth in it?" By early December he was ready to commit: "The thing with the quantum gas turns out to be very interesting," he wrote again to Ehrenfest. "I am increasingly convinced that very much of what is true and deep is lurking behind it. I am happily looking forward to the moment when I can quarrel with you about it." So what did Einstein realize about Bose's method that makes its implications so interesting and deep? How can something as mundane as a statistical counting method lead to a revolution in our physical worldview?

Any serious gambler knows that the laws of statistics are laws of nature, just as surely as is gravitational attraction. Games of chance are based on systems that are chaotic and unpredictable, such as a ball bouncing around on a rotating roulette wheel, or a pair of dice flung forcefully onto a surface. Since each toss is slightly different, and the final resting position of the dice depends sensitively on the small details of each throw, these events are effectively random processes, in which the probability that each face will turn up is the same, and equal to one-sixth. Moreover, what face turns up on one of the dice is completely independent of what face turns up on the other die. From these simple principles it is possible to work out the consequences of rolling a pair of dice many times, to the point where a casino can make an extremely reliable income from dice-based games.

Games of chance, such as dice or cards, are all based on the same underlying statistical principle: each specific configuration of the basic units (cards, dice, coins) is equally likely; this is exactly the same assumption as underlies the entropy concept in statistical physics. The atomic world behaves like a huge number of many-faced dice, constantly being rolled and rerolled; in fact Bose's combinatorial formula is essentially a statement about the number of states available when a huge aggregate of many-sided dice are thrown. To understand the strangeness of his answer, consider the simple case of throwing two dice. The available configurations are naturally specified by a pair of numbers, the number facing upward on die one and the number facing upwards on die two; for example $(1, 4)$ is a specific configuration in which die one shows a 1 and die two shows a 4. Each of the thirty-six possible pairs $[(1, 1), (1, 2), (2, 1), \ldots (5, 6), (6, 6)]$ is then equally likely to occur. However, the statistics gets somewhat more interesting when one looks, not at a specific configuration, but at the total *score* in a throw, the sum of the two numbers defining a configuration. Now one quickly realizes that there are six configurations adding up to seven (i.e., six ways to roll a seven) and there is only one way to roll a two. Thus the chance of rolling a seven is $6/36 = 1/6$, and of rolling a two is $1/36$. These calculations, and all other statistical properties of dice, follow directly from the fact that there are two distinct, independent dice, each of which randomly shows one of its faces when thrown, and that each throw is independent.

Now, if you have a pair of different-color dice (e.g., die one red, die two blue) and you keep track as you roll many times, you will surely find that (red = 3, blue = 4) and (red = 4, blue = 3) occur roughly an equal number of times, and you can tell that some of your sevens come from (3, 4) and some from (4, 3). However, suppose someone makes for you a pair of dice so perfectly matched that they are completely identical to your eye, and you put the dice in a closed box and shake them before making the throw. In this case every time you get a four and a three you will not be able to tell whether it is (3, 4) or (4, 3). Do you expect this to make any difference in the probability of getting a seven? Absolutely not. This probability is a law of physics: there are two distinct, independent physical possibilities, which the laws of dynamics may or may not lead to in a given roll, and we must *add* the probabilities for each to occur to get the right answer. It matters not at all if we can *tell* which possibility actually occurred.

What, then, about the behavior of two atoms (or electrons) being distributed by some complex microscopic dynamics into, say, six different quantum energy levels? The two atoms are then like two "quantum dice," and the energy level each atom occupies is analogous to the face of the die that comes up. If atoms are independent, distinct objects, no matter how much they look identical, one would have to conclude that having atom one in level three, and atom two in level four, is a different possibility from atom one in four and atom two in three. And therefore that these two possibilities must both contribute to the number of possible states (i.e., both contribute to the entropy of the system). One would be wrong.

This is the mind-bending, if unappreciated, assumption behind Bose's method of counting light quanta, which Einstein adopted for atoms and which he must have fully grasped only sometime after his first paper on the atomic ideal gas. The new principle is that, *in the atomic realm, the interchanging of the role of two identical particles does not lead to a distinct physical state.* This has nothing to do with whether a physicist chooses to regard these states as the same, or doesn't *know* how to distinguish them: they *are not* distinct. This is an ontological and not an epistemological assertion.

How do we know this? Consider again our quantum dice. According to Bose-Einstein statistics, there are now only twenty-one possible configurations, not thirty-six. The six doubles are still there as before [(1, 1), (2, 2) . . .]. The number of these states didn't change when we switched over to quantum dice; even with classical statistics there is only one way to get snake eyes, or double deuces, etcetera. But now, for the thirty other configurations, where the two numbers are different, we identify them pairwise, leaving only fifteen. Configurations (3, 4) and (4, 3) are merged into a single entity of "three-four-and-four-three-ness," and similarly for all the other unlike pairs. Now, suddenly, our dice behave differently. Instead of seven being the most likely score, six, seven, and eight are all equally likely and have probability 1/7 of occurring. (With the new rules one might be tempted to sneak a pair of quantum dice into a classical casino and make a killing.)

But there is a further change in the probabilities, which has a profound significance in physics. With the Bose-Einstein approach, the probability of rolling doubles has greatly increased. Classically the chance of rolling doubles is 6/36 = 1/6 = 16.6 percent; switching to the quantum dice makes it 6/21 = 28.5 percent, increasing the odds of doubles by more than 70 percent. With Bose-Einstein statistics there are fewer configurations available in which the particles do different things, and as a result the particles have a tendency to bunch together in the same states! And the more particles there are, the more there is the tendency to bunch. For three quantum "dice" the probability of rolling triples is more than twice as large as it would be if the classical statistics of distinguishable dice held sway. With a trillion trillion quantum particles, as in a mole of gas, this effect is enormous; it literally changes the behavior of matter.

Fine, but do we really care that much about what happens when you swap atoms? Well, we should. Because it is very hard to think of atoms as particles in our usual everyday sense when they lack this individuality. After all, just as we could imagine painting one die red and the other blue (i.e., labeling them), can't we somehow label atom one and atom two, and distinguish them? No, we can't (according to Einstein). Atoms are fundamentally indistinguishable and impossible to label. Nature is such that they are not separate *entities*, with their own independent trajectories through space and time. They exist in an eerie, fuzzy state of oneness when aggregated. So the Bose-Einstein statistical worldview, coming from a different direction, reinforces the concept of wave-particle duality, in this case applied to both light *and* matter, and heralds the emerging discovery that the microscopic world exists in a bizarre mixture of potentiality and actuality.

Einstein lays out this revolutionary idea in his second paper, read to the Prussian Academy on January 8, 1925, where he also predicts a totally unexpected condensation phenomenon that would have a profound influence on quantum physics up to the present. He introduces the new paper as follows: "When the Bose derivation of Planck's radiation formula is taken seriously, then one is not permitted to ignore it as a theory of the ideal gas; when it is correctly applied, the radiation is recognized as a gas of quanta, so the analogy between the gas of quanta and the gas of molecules must be a complete one. In the following, the earlier development will be supplemented by something new, which seems to me to increase the interest of the subject."

The interesting "something new" is first presented as a mathematical paradox. In his first paper he derived an equation relating the density of the quantum ideal gas to the temperature of the gas. Upon close inspection one notices an odd feature of this equation. On the left-hand side of the equation sits the density of the gas in a container, a quantity that can be increased indefinitely simply by compressing the volume of the container, which is kept at a fixed temperature.[4] But on the right-hand side is a mathematical expression that varies with temperature but cannot get larger than a certain maximum value if the

---

[4] The gas can be kept at a fixed temperature as its density (and hence its pressure) increases simply by making the container out of a heat-conducting material and putting it in contact with a large "bath" that will add or subtract the necessary heat.

temperature is fixed. This leads to an apparent contradiction, as Einstein points out: this equation violates the "self-evident requirement that the volume and temperature of an amount of gas can be given arbitrarily." What happens, he asks, when at fixed temperature one lets the density increase by compressing it into a smaller volume until the density becomes greater than the maximum allowed?

Having posed the question, he brilliantly resolves it with a bold hypothesis: "I maintain that in this case . . . an increasing number of the molecules go into the quantum state numbered 1 [the ground state], the state without kinetic energy. . . , a separation occurs; a part [of the gas] 'condenses,' the rest remains a 'saturated ideal gas.'" Here he is making an analogy to an ordinary gas, like water vapor, which when cooled reaches a temperature at which it begins to condense partly into a liquid while still retaining a particular ratio of liquid to vapor.[5] The reason that his hypothesis resolves the paradox is that in deriving the relation of density to pressure in his original paper he made an innocent mathematical transformation, which amounted to neglecting the single quantum state of the gas where each molecule has zero energy.[6] Never before in the history of statistical physics had the neglect of a single state made any difference to the value of a thermodynamic property of a gas, such as its density. On the contrary, the number of states involved, as we saw earlier, is normally unimaginably large, and physicists routinely make approximations that neglect billions of states without giving it a second thought. But Einstein unerringly recognized that in this new world of Bose-Einstein statistics, this single zero-energy state would gobble up a macroscopic fraction of all the molecules, creating a novel quantum "liquid," now known as a Bose-Einstein condensate.

---

[5] This state of an ordinary gas-liquid is referred to as "phase-separated," and the gas is termed "saturated," hence Einstein's use of the term in this new context.

[6] Quantum mavens might worry that having a particle at rest in a finite volume violates the uncertainty principle. They would be right; "zero energy" here means an essentially infinitesimal energy, much less than the natural thermal energy scale kT, determined by the macroscopic size of the container for the gas.

The generosity of the "Bose-Einstein" designation is not widely appreciated, as few physicists realize that Bose said not a word about the quantum ideal gas in his seminal paper. The paper that does predict quantum condensation belongs to Einstein alone, and it is a masterwork. The boldness of the young rebel combines with the technical virtuosity of the mature creator of general relativity to reach breathtaking conclusions with complete self-assurance. A lesser physicist would either not have noticed the subtle mathematical error introduced by the neglect of a single state or, even if noticed, would likely have dismissed its logical implications as so bizarre as to indicate some fundamental error. The reason that this condensation phenomenon seems so strange, even today, is that condensation of an ordinary gas is caused by the weak attraction between the gas molecules, which becomes important only when the gas is relatively dense. But Einstein is considering the theory of the *ideal* gas, in which such molecular interactions are assumed to be completely absent. *His* condensation phenomenon is driven purely by the newly discovered quantum "oneness" of identical particles, not by a force like electromagnetism, but by this strange statistical "pseudoforce" that Einstein was the first to recognize. Proposing it as a real physical phenomenon was an act of great courage.

Bose-Einstein condensation is now one of the fundamental pillars of condensed-matter physics; it underlies the phenomena of superconductivity of solids and superfluidity of liquids such as helium at low temperatures,[7] which have been the subject of five Nobel prizes. These substances have substantial interaction forces between atoms and electrons, unlike the ideal gas of Einstein's theory, although it is clear theoretically that the "statistical attraction" of bosonic particles plays the key role in generating their unique properties. Nonetheless, it was big news and Nobel-worthy yet again when, in 1995, atomic physicists finally realized a holy grail of the field. They created an atomic gas with

---

[7] Superconductivity is a phenomenon whereby, at low temperatures, a solid such as aluminum conducts electricity with no resistance and hence no dissipation of energy; superfluidity is a similar effect in liquids, whereby they lose their viscosity and flow without dissipating energy.

negligible interactions, cold enough[8] to observe pure Bose-Einstein condensation—Einstein's last great experimental prediction, coming to fruition a lifetime after its first statement.

And again, as before, Einstein's progress was a step too far for the physics world; it was ignored by the leading atomic physicists, such as Bohr, Sommerfeld, and Max Born, and by the soon-to-be-famous Werner Heisenberg and Wolfgang Pauli. Even Einstein's admired colleague Planck and his great friend Ehrenfest, the former student of Boltzmann, thought it was unacceptable; they believed that the Bose-Einstein statistical method was just plain wrong. In the condensate paper Einstein responds explicitly to their criticism: "Ehrenfest and others have reported that in the Bose theory of radiation and in my [ideal gas theory], the quanta or molecules do not act in a manner statistically independent of each other . . . this is entirely correct. . . . The formula [for counting states according to Bose and Einstein] expresses indirectly an implicit hypothesis about the mutual influence of the molecules of a totally new and mysterious kind." That's it. Statistical independence (which here means the classical method of counting configurations) is out; statistical attraction is in. Quantum particles bunch; get used to it.

Einstein then goes on to show that *only* Bose-Einstein statistics can save two things statistical physicists hold dear: Nernst's Third Law of thermodynamics and the additivity of entropy (i.e., when one combines two systems their total entropy is the sum of their individual entropies), a less obvious but fundamental requirement for thermodynamics to make sense. Having carried physics to the very brink of the new quantum theory, he rests his case.

While Einstein's breakthrough did not have much impact on the mainstream of quantum research, a lesser-known but already well-respected physicist had been following carefully Einstein's papers on the quantum ideal gas, the Austrian Erwin Schrödinger, who had

---

[8] The required temperature in these experiments was 170 *billionths* of a degree above absolute zero, by far the coldest temperature ever created by man at that time.

Berlin. 28.II.25

Verehrter Herr Kollege!

Erst heute komme ich dazu, auf Ihren Brief vom 5.II zu antworten. Ihr Vorwurf ist nicht ungerechtfertigt, wenn unser ein Fehler in meiner Abhandlung nicht vorliegt. In der von mir verwendeten Bose'schen Statistik werden die Quanten bezw. Moleküle nicht (als) voneinander unabhängig behandelt. Darauf beruht es, dass die Formel

$$W_2 = \frac{(n_s)^2 e^{-n_s}}{2!}$$

nicht gilt. Ich versäumte es, deutlich hervorzuheben, dass hier eine besondere Statistik angewendet ist, die vorerst nichts anderes als durch den Erfolg vorläufig begründet werden kann:

Die Komplexion ist charakterisiert durch Angabe der Zahl der Moleküle, welche in jeder einzelnen Zelle vorhanden ist. Die Zahl der so definierten Komplexionen soll für die Entropie massgebend sein. Bei diesem Verfahren erscheinen die Moleküle nicht als voneinander unabhängig lokalisiert, sondern sie haben eine Vorliebe, mit einem andern Molekül zusammen in derselben Zelle zu sitzen. Man kann sich das an kleinen Zahlen leicht vergegenwärtigen. Z. b. 2 Quanten, 2 Zellen:

	Bose – Statistik				unabhängige Moleküle	
	1.Zelle	2. Zelle			1.Zelle	2. Zelle
1. Fall	••	–		1. Fall	I II	–
2. Fall	•	•		2. Fall	I	II
3. Fall	–	••		3. Fall	II	I
				4. Fall	–	I II

Nach Bose hocken die Moleküle relativ häufiger zusammen als nach der Hypothese der statistischen Unabhängigkeit der Moleküle.

FIGURE 25.1. First page of Einstein's letter to Schrödinger of February 28, 1925 in which he explains how Bose statistics differs from classical statistics, using the diagram at the bottom of the page. The example is equivalent to two quantum coins: on the right Einstein lists the four states of two ordinary coins, whereas on the left he lists the three states of quantum coins, since heads-tails and tails-head are indistinguishable quantum mechanically. Austrian Central Library for Physics.

recently been appointed professor of theoretical physics at the University of Zurich (Einstein's first academic position). Schrödinger had previously worked on both general relativity theory and on the quantum ideal gas, and he had met Einstein recently in Innsbruck, shortly after his first paper on the ideal gas had been published. He wrote to Einstein in February of 1925, when he did not yet know of Einstein's landmark second paper on condensation, expressing skepticism about the validity of his application of Bose's method to atoms in his first paper. Einstein replied a few weeks later with characteristic good humor: "your reproach is not unjustified, although I have not made a mistake in my paper. . . . In the Bose statistics, which I use, the quanta or molecules are not considered as being mutually independent objects." He then drew a small diagram illustrating the case of just two particles in two states, pointing out that there is only one configuration in which the particles are in different states instead of two, as there would be in classical statistics. To Schrödinger, Einstein's reply was a revelation: "only through your letter did the uniqueness and originality of your statistical method of calculation become clear to me," he wrote in November of 1925. "I had not grasped it before at all despite the fact that Bose's paper had already come out. . . . [Bose's work] did not seem particularly interesting to me. Only your theory of gas degeneracy is really something fundamentally new."

Schrödinger wanted to follow up on the suggestion, which Einstein had made at the end of the paper, that the strange statistics could be understood using the concept of matter waves, recently introduced by the young French physicist Louis de Broglie. Within months this newly inspired research direction would fundamentally change the emerging modern form of quantum theory, and enshrine Herr Schrödinger in the eternal pantheon of physics.

CHAPTER 26

THE ROYAL MARRIAGE:
$E = mc^2 = h\upsilon$

I said to myself that classical physics wasn't sufficient, that all of the ancient edifice . . . was shaken and that it was necessary to reconstruct the edifice; but I didn't think it was necessary to change fundamental notions, I thought that it was by the introduction of new things, completely unknown, that one would take into account quanta. . . . For example by a synthesis between waves and particles . . . it was more or less the point of view of Einstein.
—LOUIS DE BROGLIE

"A younger brother of the de Broglie known to us has made a very interesting attempt to interpret the Bohr-Sommerfeld quantization rules in his dissertation. I believe that it is the first feeble ray of light to illuminate this, the worst of our physical riddles. I have also discovered something that supports his construction." So Einstein wrote to Lorentz in December of 1924, just as he was completing his masterpiece on the quantum ideal gas. In the midst of his cogitations on the meaning of Bose statistics, during the summer of 1924, a second bolt from the blue singed his mail slot. This revelation came in the form of a doctoral thesis, sent to him by his old friend Paul Langevin, the French physicist, with the somewhat skeptical assessment, "[the thesis] is a bit strange, but after all, Bohr also was a bit strange, so see if it is worth something all the same." Just as with Bose, Einstein was willing to look at this young man's ideas with an open mind and recognize the crucial insight they represented. He wrote back to Langevin with a warm and eloquent endorsement: "Louis de Broglie's work has greatly impressed me. *He has lifted a corner of the great veil.* In my work I [have recently] obtained results that seem to confirm his. If you see him please tell him how much esteem and sympathy I have for him."

Louis de Broglie, like Bose, had venerated Einstein from his earliest exposure to modern physics, but, unlike Bose, he had good reason to expect that Einstein might someday take his ideas seriously. Louis' older brother, Maurice, was one of the most distinguished experimental physicists in France and had been one of two scientific secretaries at the First Solvay Congress of 1911 (hence Einstein's allusion to the "de Broglie known to us" in his letter to Lorentz). Moreover, the de Broglies were an eminent noble family of France; Maurice and Louis counted among their predecessors ministers, generals, and famous literary figures. Maurice himself held the title of Duke de Broglie, whereas Louis had the inherited rank of Prince of the Holy Roman Empire, awarded to all the direct descendants of his ancestor Duke Victor-François in reward for his martial feats, by order of the emperor Francis I.

Louis Victor Pierre Raymond de Broglie was born on August 15, 1892, the youngest of the five offspring of his father Victor de Broglie. His brother, Maurice, was the second child, seventeen years older than Louis and first son (hence duke). He served as a father figure and mentor to Louis, particularly after the death of Victor in 1906; "at every stage of my life and career," Louis wrote to his brother, "I found you near me as guide and support." Maurice had begun a career as a naval officer in 1895 but, against the wishes of his family, abandoned this path in 1904 and devoted himself to science, taking the rather unusual step of installing a private laboratory in the family town house on the tony Rue Chateaubriand, adjacent to the Arc de Triomphe. He would become an enormously influential figure in French physics, successor to Langevin at the College de France, and nominated for the Nobel Prize multiple times, including in the same year, 1925, when his younger brother was first nominated.[1]

---

[1] Unlike Louis, Maurice was never awarded the prize, although his research was prominently cited in the 1922 prize, awarded to Niels Bohr.

Louis' sister and close companion Pauline described the young prince glowingly:

> this little brother had become a charming child, slender, svelte, with a small laughing face, eyes shining with mischief, curled like a poodle. . . . His gaiety filled the house. He talked all the time, even at the dinner table, where the most severe injunctions to silence could not make him hold his tongue. . . . He had a prodigious memory . . . and seemed to have a particular taste for . . . political history. . . . He improvised speeches inspired by accounts of the newspapers and could recite unerringly complete lists of the Ministers of the Third Republic. . . . A great future as a statesman was predicted for Louis.

After finishing his secondary education in 1909, graduating from the elite Lycée Janson de Sailly, and obtaining his *licence* in historical studies, he was encouraged to "continue and prepare the diploma in history." However, this early passion was dying out, and confusion about his future set in. "I could see that to do that it was necessary to go often to the library, make a large bibliography and such things. That didn't appeal to me too much, and that was the beginning of the year of my 'moral crisis.'" He studied law for a year while hesitating "between several intellectual directions . . . and [I] wasn't very much in agreement with my brother, who would have preferred that I do either the Ecole Polytechnique or study diplomacy. . . . I wanted to do neither, and yet I didn't want to follow him [into physics], which increased the crisis a bit." Finally he committed himself to a course in advanced mathematics, and by the end of the year, according to his brother, "The hesitations are over, he has crossed the Rubicon, and the course of his thoughts has turned towards physics and more particularly theoretical physics."

The precise timing of this crisis and conversion is not clear, but it overlapped closely with Maurice's service as secretary at the First Solvay Congress, which afforded Louis a unique window into the ferment of the new atomic theory. De Broglie recalled: "I began to think about quanta from the moment that my brother gave me the notes of the Solvay Congress of 1911, probably at the beginning of 1912," and "with the ardor of my age I became enthusiastically interested in the problems that had been treated and I promised myself to devote all my efforts to achieve an understanding of the mysterious quanta, which Max Planck had introduced ten years earlier into theoretical physics, but whose deep significance had not yet been grasped." Two years later, in 1913, he graduated with his *licence de science*, having performed brilliantly on his exams. "His enthusiasm was returning," Maurice recalled, "with the certainty of being at last on the right track."

However, just as the young man was on the verge of entering the intellectual fray of quantum research, all of Europe was engulfed in a more primitive conflict, which led to an indefinite extension of the brief military service to which de Broglie had been called after graduation. He was eventually assigned to a wireless telegraphy unit and found himself working *underneath* the Eiffel Tower, where a transmitter had been installed. And while his brother recalled that Louis "regretted the interruption of his meditations" on quantum theory and complained of "inspiration broken to pieces," Louis himself spoke of the experience somewhat fondly, "because that put me in relation with . . . real things. That made a certain impression, and led me to examine quite closely all the theories of the propagation of waves, all the theories of electronics . . . that certainly contributed greatly to clarifying my ideas on all this." Decommissioned in 1919, de Broglie was ready to begin his quest in earnest.

De Broglie was determined to be a theorist, unlike his brother, but this was not nearly as felicitous a choice in France as in the German-speaking countries of Europe, which were leading the abstract developments of quantum theory. De Broglie's close friend and fellow theorist Leon Brillouin recalled, "There was really no career open for a theoretical physicist in the French organization. People who had curiosity for theory would go right away into pure mathematics. . . . Many of my colleagues told me 'Are you crazy? To go into theoretical physics, there is no future." Nonetheless Brillouin, and de Broglie, joined the small group around Paul Langevin who were working on quantum theory.

De Broglie's first thought was to focus on the theory of light quanta, which, unlike most other physicists, he had believed in since his first exposure to Einstein's work in 1912.

> I never had any doubt at this time about the existence of photons. I considered that Einstein had discovered them, that they raised many difficulties . . . but that in the end this was a problem to resolve, that one could not deny the existence of photons. . . . It must be noted that I was very young. I had not made any theoretical works and for that reason was not as attached to electromagnetic theory, as was Langevin. . . . Thus, I accepted this as something that must be required.

In addition, his brother Maurice had been very directly grappling with the particulate properties of light in his extensive research on the x-ray photoelectric effect. Louis and Maurice collaborated on the interpretation of these experiments in terms of Bohr's theory during 1920 and 1921; particularly striking was the fact that the x-ray radiation appeared capable of giving all its energy to a pointlike electron, an observation very difficult to reconcile with a picture of the x-rays as extended, spherically expanding waves. When Maurice spoke of his experiments at the 1921 Solvay Congress (which Einstein had skipped), he concluded that the radiation "must be corpuscular . . . or if undulatory, its energy must be concentrated in points on the surface of the wave."

De Broglie's first significant contribution had a motivation remarkably similar to that of Bose: to derive "a number of known results of the theory of radiation . . . without the intervention of electromagnetic theory" (i.e., without using classical physics). "The hypothesis we adopt is that of light quanta." However, unlike every other physicist working on the quantum problem, de Broglie was convinced that the key to unlocking it lay in the appropriate use of relativity theory, which he described as Einstein's "incomparable insight." Thus, in the second paragraph of the paper, he introduces a very peculiar notion, that quanta are to be thought of as "atoms" of light, with a very small but nonzero mass. In almost two decades of quantum research neither Einstein nor anyone else in the field had made such an outré suggestion.

Photons (quanta) were conceived to have energy, $E = hv$, and momentum, $p = hv/c$, but their energy and momentum was not accompanied by a rest mass, precisely because photons, by definition, move at the speed of light and can never be at rest in *any* frame of reference. Nonetheless de Broglie asserts that photons *should* be conceived of as having a rest mass which satisfies Einstein's most famous equation, $E = mc^2$! Hence, he states, combining this with the Planck relation, $E = hv$, this mass must equal hv/c^2. Never mind that this mass would then vary continuously with the frequency of light, which seemed very odd—in the current work he simply assumes that this mass is "infinitely small," while the speed of the photon is "infinitely close" to c, so that the usual relation $E/c = p$ for photons still holds, to a very good approximation.[2]

It must be noted that in the modern quantum theory the mass of the photon *is* precisely zero, and a finite photon mass played no role in the first full formulations of quantum mechanics, which emerged a mere four years later. Thus this forced marriage of the two most famous equations of modern physics ($E = mc^2$ and $E = hv$) ended in a speedy divorce. However, in the first paper de Broglie uses this hypothesis only to rederive a known result of electromagnetic theory, that light exerts pressure and that this pressure is half the value one finds for ordinary nonrelativistic particles.[3] From this point on in the paper he uses standard statistical and thermodynamic relations, very much as in Einstein's work of 1902–1906, to derive the law for blackbody radiation, arriving not at Planck's formula but rather at Wien's law.[4] The reason for this is that he has, like so many before, assumed that the quanta are statistically independent, and, as was discussed in connection with the work of Bose, this assumption inevitably leads to Wien's approximation to the Planck law. He misses completely the strange statistical at-

---

[2] Note that c here can no longer be thought of as *the* speed of light, but rather the limiting velocity of light as its frequency and hence mass goes to zero.

[3] When measured appropriately, in terms of its energy density.

[4] De Broglie mentions in passing that if one were to consider not just isolated atoms of light but "a mixture of monatomic, diatomic, triatomic" molecules of light, Planck's law could be obtained, but then dismisses this as requiring "some arbitrary hypotheses." He and others followed up this idea, but it was superseded by the concepts of Bose statistics.

traction that Bose stumbled upon, and Einstein would elucidate, two years later. The paper is significant for two reasons. First, de Broglie introduces the method of counting single photon states using units defined by Planck's constant, which Bose would rediscover a couple of years later.[5] Second, de Broglie expressly assumes that the formulas of relativity theory are used for "atoms of light" and thus asserts that relativistic mechanics is of central importance in quantum theory, whereas previously it played a very minor role. This novel point of view would shortly lead him to a historic breakthrough.

After his initial foray into the theory of light quanta in 1922, de Broglie became convinced that there must be some symmetry between the behavior of "atoms of light" and other massive particles, such as electrons or atoms. According to his friend Brillouin, he was fascinated by the experimental images of radioactive decay processes, in which massive particles (e.g., electrons and positrons) are emitted from the atomic nucleus and follow curved tracks due to the force exerted by a magnetic field, while the simultaneously emitted photon makes a straight track, since it lacks electric charge. Apparently de Broglie intuitively felt, "Well, all this must be very similar. Either they are all waves or they are all particles . . . [so he tried] to see if he couldn't make everything waves." De Broglie recalled that the key ideas "developed rapidly in the summer of 1923," perhaps in July. "I got the idea that one had to extend [wave-particle] duality to the material particles, especially to electrons." In September and October of that year he submitted three short notes for publication, which contained "the essential things" that later entered into his thesis.

In pondering how to associate a wave with material particles, de Broglie had been struggling with an apparent paradox. Just as had he had for photons, he persisted in combining the two great equations of Einstein and Planck and hence associating a frequency with the particle's rest mass: $m = h\nu/c^2$, but now he took the bold step of applying

---

[5] Like Bose, de Broglie then finds his answer is off by a factor of 2, and he needs to insert this factor "by hand" to account for the two possible polarizations of light (which is a concept of classical electromagnetism, not present in the theory of light quanta at that time).

this formula to electrons, for which the mass was not unmeasurably small. De Broglie insisted that this was "a meta law of Nature, [that] to each . . . proper mass m, one may associate a periodic phenomenon of frequency $v = mc^2/h$." In other words de Broglie postulated a sort of internal "vibration" of every particle, which acted like a ticking clock, even when it was at rest. Moreover, if this is so, reasoned de Broglie, then when the particle moves at velocity v, its "ticking" must slow down, because Einstein's theory of relativity predicts the universal effect of *time dilation*: clocks are measured to run more slowly when in motion relative to an observer. So the moving particle will appear to vibrate at a lower frequency,[6] v_1, than it does when at rest.

However, de Broglie at the same time considered the most basic postulate of relativity theory, that the laws of physics are the same in all frames of reference, and hence that the energy of the moving particle must still be related to *some* frequency via Planck's constant (i.e., $E = hv$ must still hold when the particle is moving). But since the energy of the particle is *larger* in the frame in which it is moving (it has kinetic energy, $\frac{1}{2}mv^2$, in addition to its rest mass energy), then to satisfy the Planck relation it must have a *higher* frequency,[7] v_2, than its "rest frequency," v. So there were two frequencies one should associate with the particle motion, one *larger* than its rest frequency, one *smaller*; which one was physically relevant?

Both, was de Broglie's answer. While the particle is moving, "it glides on its wave, so that the internal vibration of the particle (v_1) is in phase with the vibration of the wave (v_2), at the point where it finds itself." The "fictive wave" that guides the particle moves at just the right speed so that the wave peak coincides with the peak of the particle oscillations; the particle is like a lucky surfer, permanently attached to the crest of the perfect wave. De Broglie showed that, for this to be so, the velocity of the "phase wave" (as he termed his fictive waves) must have a specific value, $V_{phase} = c^2/v$, which is faster than the speed

[6] According to relativity theory, $v_1 = v_0(1 - v^2/c^2)^{1/2}$.
[7] Again from relativity theory, $v_2 = v_0(1 - v^2/c^2)^{-1/2}$, so it is higher than v_0 by just the same factor that v_1 is lower.

of light.[8] Because the phase waves moved faster than light, they could carry no energy, according to relativity theory, but served only to guide the particle motion.

So that is de Broglie's picture: every particle has some unspecified internal oscillation, which must remain in phase with a mysterious steering wave that directs its motion but which moves ahead faster than light so as to remain always "in resonance" with the particle's oscillation. Even by the standards of the new quantum theory, this was a rather wild invention, and, if anything, Langevin greatly understated the case when he told Einstein it was "a bit strange."

But de Broglie had at least one further result that supported his extreme conjectures. He took this picture and applied it to an electron circulating around a hydrogen atom. The electron would move around in a circular orbit, continually emitting these phase waves, which would zoom ahead of the particle and lap the particle almost 19,000 times for each electronic circuit. Again he asked the question: what was required so that each time a wave crest and the particle coincided, the particle oscillation and the wave oscillation were in phase? Almost miraculously, he was able to show that this requirement was equivalent to the Bohr-Sommerfeld rule for the allowed electron orbits in hydrogen.[9] It was this result that apparently impressed Einstein, who referred to it, in his paper on Bose-Einstein condensation, as "a very remarkable geometric interpretation of the Bohr-Sommerfeld quantization rule."

While de Broglie's key ideas were developed in the fall of 1923, he prepared a longer and more comprehensive document as his thesis in the spring of 1924, and gave it to his adviser, Langevin, who clearly was concerned about whether by accepting the work he would be endorsing nonsense. "It looks far-fetched to me," was his initial reaction, shared with a colleague. At some point in the spring Langevin spoke to Einstein, who was intrigued and agreed to look at a copy. Einstein "read my thesis during the summer of 1924" de Broglie recalled, and

[8] Since any massive particle's velocity must be less than the speed of light, the wave velocity, $V_{phase} = (c/v)c$, is necessarily greater than the speed of light.

[9] Suitably generalized to include relativistic effects.

wrote the very favorable report quoted above. "As M. Langevin had great regard for Einstein, he counted this opinion greatly, and this changed a bit his opinion with regard to my thesis."

De Broglie then defended his thesis in November of 1924 before a distinguished but bewildered "jury" (as it is termed in France), chaired by the future Nobel laureate Jean Perrin and including the famous mathematician Elie Cartan, the eminent crystallographer Charles Mauguin, and his adviser, Langevin. While their verdict was positive and de Broglie was congratulated for his "remarkable mastery," a student who attended the thesis defense later remarked, "never has so much gone over the heads of so many." Maurice de Broglie sought a candid opinion from Perrin and was told, "all I can tell you is that your brother is very intelligent."

De Broglie's thesis had come to Einstein's attention at the perfect time. Einstein was now deep into his second paper analyzing the statistical properties of the quantum atomic gas. And in addition to his realization that the principle of indistinguishability of particles is implicit in Bose statistics, and the possibility of quantum condensation this implied, he had made one more major mathematical discovery. Just as he had done for the gas of light quanta, which he had analyzed in the seminal work of 1909, he now looked at the *fluctuations* of the energy in a particular volume of the quantum gas of particles. For both the photon gas and for a gas of atoms in a box, the energy in a small region of the box can vary randomly in time, while maintaining the same energy on average. This is simply a reflection of the ceaseless give-and-take that corresponds to thermal equilibrium. Einstein's earliest insight into the wave-particle duality of light had come in 1909, when he derived a formula for the typical magnitude of these energy fluctuations. He found that it consisted of two contributions, one of which could be explained by the interference of light waves, but the other of which looked exactly like the fluctuations expected from a gas of particles with energy $E = h\nu$. It was this latter, particulate term that was a revelation in 1909, supporting Einstein's hypothesis of light quanta and prompting him to declare that the future quantum theory would involve a "fusion" of the wave and particle concepts. Now, after adopting Bose statistics, he finally had the correct theory to evaluate the same quantity for the atomic gas. He found that *exactly the same structure occurs*: the fluctuations are the sum of two terms, a "particle term" and a "wave term." But in this case it is the "wave term" that is the surprise.

This parallelism between atoms and light impressed Einstein greatly because, as he says in his second quantum gas paper, "I believe that it is more than just an analogy, since a material particle . . . can be represented by a . . . wave field, as de Broglie has stated in a remarkable thesis." He continues: "this oscillating field—whose physical nature is still obscure—must, in principle permit itself to be demonstrated by phenomena corresponding to its motion. So a beam of gas molecules travelling through an opening must experience a bending, which is analogous to that of a beam of light." That was the key to testing this extension of wave-particle duality. Whereas before, one was looking for *particulate* behavior of light waves, such as the localized collisions of photons and atoms in the photoelectric effect, now one should look for *wave* behavior of particles: a stream of atoms should interfere with itself as it passed near an edge, exhibiting the wave phenomenon known as diffraction. Einstein began suggesting such a search to physicists in September of 1924, shortly after reading de Broglie's thesis.[10]

Einstein's endorsement and extension of de Broglie's matter wave concept was decisive in bringing this idea under serious study in the physics community. "The scientific world of the time hung on every one of Einstein's words, for he was then at the peak of his fame," de Broglie noted. "By stressing the importance of wave mechanics, the illustrious scientist had done a great deal to hasten its development. Without his paper my thesis might not have been appreciated until very much later." Even *with* his paper the thesis was viewed with suspicion; a student of Sommerfeld's recalled, "the paper of de Broglie was studied [in Munich] too; everyone had objections (they were not very difficult to find), and no one took the idea seriously." Thus the search to observe matter waves, which began almost immediately after Einstein's second paper on the quantum gas, was mainly pursued in Einstein's name.

[10] De Broglie also had suggested such a search, for interference of electrons, roughly a year earlier.

Walter Elsasser, the very first experimenter to find evidence in this direction, illustrates this with the introduction to his initial paper: "By way of a detour through statistical mechanics, Einstein has recently arrived at a physically very remarkable result. Namely he makes plausible the assumption that a wave field is to be associated with every translational motion of a material particle. . . . The hypothesis of such waves, already advanced by de Broglie before Einstein, is so strongly supported by Einstein's theory that it seems appropriate to look for experimental tests for it." Within two years these tests rendered an unambiguous verdict: electrons beams do show wavelike interference. This discovery was so striking that de Broglie, with Einstein's strong support, received the Nobel Prize in Physics in 1929, a mere four years after his work became widely known.

But had this enormous success corroborated de Broglie's specific model, involving superluminal phase waves? Not at all. Almost no trace of this concept survives in modern quantum mechanics, and the vast majority of contemporary physicists are completely unaware of the basis on which de Broglie argued for matter waves. The only equation that survives from de Broglie's thesis work is so simple that Einstein could have written it down any time after 1905.[11] This is the famous relation $\lambda = h/p = h/mv$, which relates the momentum, $p = mv$, of a massive particle to its "De Broglie wavelength." It is obtained in a few lines by an extension, from light quanta to material particles, of Einstein's equation $E = hv$.[12] This formula, correctly interpreted, is essential to modern quantum theory but arises there without any appeal to relativity theory or de Broglie's phase waves.

---

[11] Much later, under direct questioning from the physicist I. I. Rabi, Einstein allowed that he did indeed think of the famous equation $\lambda = h/p$ for matter waves before de Broglie but didn't publish because "there was no experimental evidence" for it.

[12] The logic is as follows: for a photon, $E = hv$, and for light waves $E = pc$. If we assume both relations hold and use the relationship of wave frequency to wavelength, $v = c/\lambda$, we get $\lambda = h/p$. If we assume the same relation holds for massive particles moving slowly compared with the speed of light, so that $p = mv$, we find $\lambda = h/mv$. The full quantum derivation of this is based on Schrödinger's equation and doesn't rely on the assumptions that are used in this simple argument.

De Broglie himself, having completed his thesis by age thirty-one, never again made a fundamental contribution to physics, although he remained active in research, unlike Bose. He was an "uninspiring" classroom teacher, who started and ended his lectures precisely on time and permitted no questions during or after them. The research seminars he organized for many years were stilted affairs, with only brief open interchanges, described as "dry and devoid of passion." The disciples that congregated around him were "not of the highest intellectual caliber" and created "an atmosphere of admiration, not to say adulation." For example, it was considered bad form to refer to "quantum mechanics," even after it became the standard term; one was supposed rather to say "wave mechanics," as homage to de Broglie's seminal role. His research career spanned more than fifty years (he lived to age ninety-five), and it is now widely acknowledged that his influence was not very positive for the development of theoretical physics in France.

FIGURE 26.1. Prince Louis De Broglie circa 1930. Academie des Sciences, Paris, courtesy AIP Emilio Segrè Visual Archives.

For much of his life de Broglie worked diligently within the standard quantum theory, which emerged from the work of Schrödinger, Heisenberg, and Bohr, although he initially opposed it in 1927. Then, in 1952, at age sixty, he again rejected this approach and joined Einstein in searching for a new and more aesthetically satisfying theory. In 1954, a year before his death, Einstein wrote touchingly to de Broglie, "Yesterday I read . . . your article on quanta and determinism, and your ideas, so clear, have given me great pleasure. . . . I must resemble the bird from the desert, the ostrich, hiding its head in the sands of Relativity rather than to face the malicious Quanta. Indeed, exactly like you, I am convinced that one must look for a substructure, a necessity that the present quantum theory hides."

CHAPTER 27

THE VIENNESE POLYMATH

Physics does not consist only of atomic research, science does not consist only of physics, and life does not consist only of science.
—ERWIN SCHRÖDINGER

"When you began this work you had no idea that anything so clever would come out of it, had you?" This question was addressed to the Austrian theorist Erwin Schrödinger sometime in the fall of 1926. The questioner was a young female admirer of the thirty-nine-year-old physicist, whose unusual marriage allowed for many such "friendships." The work in question was that leading to the most famous equation of quantum mechanics, the "wave equation," named after its inventor. Schrödinger's scientific colleagues were less restrained in their praise. The reserved Planck effused, "I have read your article the way an inquisitive child listens in suspense to the solution of a puzzle which he has been bothered about for a long time." Einstein, who learned of the work from Planck, wrote simply, "the idea of your article shows real genius."

At the time of this seminal work, Schrödinger was a professor at the University of Zurich, occupying the very same chair that Einstein had once held as his first academic position.[1] Schrödinger was in the midst of what he called his "First Period of Roaming," during which

---

[1] When Einstein held the theoretical physics position is was only at the level of an associate professorship (extraordinarius); it was subsequently upgraded to a full professorship (ordinarius).

he moved between various positions, as had Einstein fifteen years earlier, ascending the academic hierarchy. Indeed, in 1927, after the great triumph of his wave equation, Schrödinger would end up as Einstein's colleague in Berlin, after receiving the signal honor of succession to the chair of the recently retired Planck. Even before that, Einstein and Schrödinger had become allies in the struggle and competition to create the new atomic theory, and they shared certain intellectual habits. Schrödinger, like Einstein, did almost all his research alone, unlike the other school of quantum theory involving Bohr, Sommerfeld, Max Born, Werner Heisenberg, Pascual Jordan, and Wolfgang Pauli, who primarily worked collaboratively. Also, Schrödinger and Einstein had a sincere respect for and interest in philosophy,[2] and they shared a similar philosophy of science, influenced by the positivism of Ernst Mach but with a strong note of idealism.

However, unlike Einstein, Schrödinger had been appointed at Zurich primarily for his breadth of knowledge, outstanding mathematical abilities, and brilliant intellect—not because of any breakthrough attached to his name. In 1926, when he finally wrote his name into the history of science, he was already thirty-nine years old, well past the age when radical breakthroughs are expected from a theoretical physicist. And in fact his style of research had never before involved a daring leap into the unknown; instead his modus operandi was to criticize and improve the work of others.

> In my scientific work . . . I have never followed one main line, . . . my work . . . is not entirely independent, since if I am to have an interest in a question, others must also have one. My word is seldom the first, but often the second, and may be inspired by a desire to contradict or to correct, but the consequent extension may turn out to be more important than the correction.

[2] Schrödinger's interest in philosophy was so great that in 1918, before the war ended, he had been planning to "devote himself to philosophy" more than physics, only to find that the chair he expected to receive, in Czernowitz, Ukraine, had disappeared along with Austrian control of the region.

In a sense, his work culminating in the wave equation *was* in that vein, building strongly on the insights of Einstein and de Broglie, but in this instance the extension was of historic consequence. In fact the state of quantum theory in 1925 called for just such an outsider, a critic who understood the two main lines of research, the Bohr-Sommerfeld atomic theory and the Einstein–Bose–De Broglie statistical theory of quanta, but who had a sentimental attachment to neither.

Erwin Schrödinger himself, while a man of great personal magnetism, was not known for his sentimental attachments. In his autobiographical sketch, written in his seventies, he reflected that he'd had only one close friend in his entire life and that he had "often been accused of flirtatiousness, instead of true friendship." Flirtatiousness understates his behavior with respect to the opposite sex. He ends his sketch with the most titillating of disclaimers. "I must refrain from drawing a complete picture of my life, as I am not good at telling stories; besides, I would have to leave out a very substantial part of the portrait, i.e. that dealing with my relationships with women." Thus we do not learn, for example, the name of the mystery woman (not his wife) who accompanied him on the Christmas ski vacation of 1925 during which the wave equation was discovered.[3]

Born in 1887 and raised in an imperial Vienna that represented the flowering of art and culture at the turn of the century, Erwin Schrödinger was closer to Einstein's generation than he was to the rising cohort of brilliant young theorists (Heisenberg, Pauli, Dirac)[4]

---

[3] While Schrödinger was discreet in this final public document, in 1933, to his diary, he confided that he never slept with a woman "who did not wish, in consequence, to live with me for all her life." There is some evidence to back this up.

[4] An English theoretical physicist, Paul Adrien Maurice Dirac, was the last of the trio of wunderkind to play a founding role in quantum mechanics, along with Pauli and Heisenberg. Born in 1902, he was even younger than Heisenberg, and upon hearing Heisenberg speak at Cambridge in July of 1925, he shortly afterward invented his own, mathematically elegant version of the quantum equations. A few years later he discovered the "Dirac Equation," the quantum wave equation that takes into account the effects of relativity. However, he interacted little with Einstein during the period 1925–26 and so is not very relevant to our historical narrative; Einstein remarked of him in August 1926, "I have trouble with Dirac. This balancing on the dizzying path between genius and madness is awful."

who would join him in driving the quantum revolution to completion. An only child, raised by a doting mother and aunts, he showed great intellectual talent from an early age. His father had studied chemistry at university, and pursued serious interests in art and botany, but contented himself with running the family linoleum business, while investing his son with his unrealized professional aspirations. Homeschooled until the age of eleven, Schrödinger then attended the elite Akademisches Gymnasium, Vienna's oldest secondary school, where he was the top student in his class for eight straight years. "I was a good student in all subjects, loved Mathematics and Physics, but also the strict logic of the ancient grammars (Latin and Greek)," he recalled. Unlike Einstein, the independent-minded Schrödinger managed to get along with his teachers and, in looking back, could "only find words of praise for my old school." His intellectual facility astonished his classmates, one of whom recounted: "I can't recall a single instance in which our Primus[5] ever could not answer a question."

When he matriculated at the University of Vienna in 1906, his brilliance was already widely known; a friend, Hans Thirring, recalls encountering a striking blond young man in the mathematics library and being told by a fellow student, sotto voce, "das ist der Schrödinger."[6] Their first meeting instilled in Thirring the conviction that "this man is really somebody special . . . a fiery spirit at work." By the time Schrödinger reached adulthood his erudition was legendary; he lectured comfortably in German, English, French, and Spanish, recited and wrote poetry (even publishing a volume late in life), and became a true expert in the philosophy of Schopenhauer and the Hindu spiritual texts, the Upanishads. Schrödinger "would translate Homer into English from the original Greek, or old Provencal poems into German," and insisted throughout his life that study of the ancient Greek thinkers was not something for his "hours of leisure" but was "justified by the hope of some gain in the understanding of modern science." It was said of Schrödinger's physics articles that "if it were not for the mathematics, they could be read with pleasure as literary essays."

---

[5] First in class.
[6] "That is *the* Schrödinger."

After settling on physics as his main focus at the end of his under-graduate years, Schrödinger went on to graduate work, primarily in experimental physics or in theoretical topics relating to the experimental work going on at the university. "I learnt to appreciate the significance of measuring. I wish there were more theoretical physicists who did." However, by the end of this period, around 1914, when he obtained his habilitation, he had decided that he was personally unsuited to be an experimenter and that Austrian experimental physics was second rate. Nonetheless he continued to do some laboratory work, and his reputation as a broadly trained physicist, conversant with both experiment and theory, would be of great value when he began searching for academic positions.

Schrödinger was poised to dive into the rushing currents of change in theoretical physics in 1914, with Bohr's atomic theory newly hatched and Einstein's general relativity on the near horizon. But, as it did for de Broglie, the Great War intervened. Schrödinger was called into service as an artillery officer, and he served in that capacity for three years before being transferred to the meteorology service. In general Schrödinger's military assignments were not among the most challenging or dangerous, and he mainly suffered from boredom, and a certain degree of depression, during this period. However, early on in his tour of duty, in October of 1915, he was caught up in one of the major battles around the Isonzo River on the Italian front and received a citation for "his fearlessness and calmness in the face of recurrent heavy enemy artillery fire."

During his war service he wrote to his many women friends, but only one visited him at the front, a young woman from Salzburg named Annemarie Bertel, whom he had met through friends in 1913. She admired and adored Schrödinger from their first meeting: "I was impressed by him because, first of all, he was very good-looking." They would marry in 1920, and within a few years the marriage evolved into a close, but nonmonogamous, relationship, with both fairly openly engaging in affairs, although Erwin was certainly the more active in this regard. For Annie (as she was known), this was the price of involvement with a great man. "I know it would be easier to live with a canary bird than with a race horse. But I prefer the race horse."

When Schrödinger returned full time to physics research in 1918, he was not particularly focused on the problems of quantum theory. He had learned theoretical physics at university from Fritz Hasenohrl, a leading disciple of the great Boltzmann, who along with Maxwell and Gibbs founded statistical mechanics. Boltzmann had died by suicide in 1906, the same year that Schrödinger began his studies; but his atomic worldview now prevailed; it had become a pillar of modern physics. "No perception in physics has ever seemed more important to me than that of Boltzmann," Schrödinger recounted, "despite Planck and Einstein."

During the war he had filled several notebooks with statistical calculations very much in the spirit of Einstein's early work on Brownian motion and diffusion. Upon returning to civilian life he published two papers based on these notes, the second of which, dealing with fluctuations in the rate of radioactive decay, is the longest article he ever produced, stretching to sixty journal pages. It was a tour de force of applied mathematics, and it announced to the world that he was to be taken seriously as a statistical physicist. In the same period he also published his first paper on quantum theory, focusing on further developments in Einstein's quantum theory of specific heat, as well as two short papers analyzing the equations of general relativity. In yet another nod to Einstein's work, in 1919, he performed an experiment trying to distinguish between the wave and particle theories of light, using a very small source. The experiment was similar in a general sense to the failed experiment that Einstein proposed in 1921 (his "monumental blunder") and gave similarly equivocal results.

Schrödinger was establishing his research style as a critic and polymath, one able to work expertly in many subfields at once, who took the ideas of others and either demolished them or clarified and extended them. Although his radiation experiment had not had been a major success, it resulted in an invitation from Sommerfeld to visit Munich, where he became enamored of the (old) quantum theory of atomic spectra, due to Bohr and elaborated in great detail by the "beautiful work" of the Sommerfeld school.[7] By 1920 he had been

---

[7] It is likely that it is at this time he studied and understood Einstein's 1917 reformulation of the Bohr-Sommerfeld theory, which he later praised so highly.

FIGURE 27.1. Erwin Schrödinger circa 1925. AIP Emilio Segrè Visual Archives, Physics Today Collection.

appointed full professor at Breslau, and he threw himself into research on atomic spectra, something Einstein had never been willing to do. By January of 1921 he had produced a step forward in the theory of alkali atoms, leading to a correspondence with Bohr, who wrote: "[your paper] interested me very much . . . some time ago I made exactly the same consideration."[8] He would continue to make respectable, but not decisive, contributions to the Bohr-Sommerfeld theory regularly, into the fateful year of 1925, when the old theory would be overthrown by two revolutions, one of his own making.

By 1922 he had been recruited to Zurich and was a certified expert in both modern quantum theory and modern statistical physics, but still a virtuoso without a masterpiece of his own. Almost all his work for the next four years would be on either atomic spectra or the statistical mechanics of gases; surprisingly, it was the latter that led him to his great discovery, with more than a nudge from Albert Einstein. As we have seen, in February of 1925, shortly after the publication of Einstein's key paper on the quantum theory of the ideal gas and Bose-Einstein condensation, Schrödinger wrote to Einstein respectfully but firmly suggesting that his paper contained an error. When, in his reply, Einstein explained to him how the new statistics worked, the scales dropped from Schrödinger's eyes, and he was entranced by the "originality of [Einstein's] statistical method." He immediately set out to deepen his understanding of this new form

---

[8] This is always a great compliment from one physicist to the other, *unless* the former is angling for priority.

of statistical physics, which he would soon describe as "a radical departure from the Boltzmann-Gibbs type of statistics."

By July of 1925 he had produced a typically insightful but incremental response, a paper titled "Remarks on the Statistical Definition of Entropy for the Ideal Gas," which contrasted Planck's definition of entropy for the gas with that of Einstein. Planck for some time had been suggesting a weaker form of indistinguishability of gas particles than that of Bose and Einstein,[9] which was sufficient to save Nernst's law but didn't lead to the weird statistical attraction that is implied by Bose-Einstein statistics. Schrödinger realized that Planck's method was illogical because it got rid of *too many* states. Recall that Bose-Einstein's new counting method, when applied to dice, would insist that the two dice "states" (4, 3) and (3, 4) are just one state, so that for each such unequal pair one should count only one state, not two, reducing the number of states and hence the entropy of the system. However, there is no such reduction for doubles (there is only one way to roll snake eyes); so there is no reduction in the number of double "states" for quantum versus classical dice. Yet Planck's method, once you understood it deeply, boiled down to counting each double as only *half* a state, which was clearly wrong. Schrödinger says exactly this: "in order that two molecules are able to exchange their roles, they must really have different roles . . . one is [then] almost automatically led to that definition of the entropy of the ideal gas which has recently been introduced by A. Einstein [Bose-Einstein statistics]." In a quaint custom of the time, this rather significant criticism of Planck was read to the Prussian Academy by Planck himself, on behalf of Schrödinger.

Einstein was impressed by this exegesis, which he himself apparently had not appreciated; in September of 1925 he wrote to Schrödinger again: "I have read with great interest your enlightening considerations on the entropy of ideal gases." He then sketched for Schrödinger another approach to the ideal gas problem, which he had

[9] For expert readers, this is the "division by $N!$" in the partition function (state counting), which is approximately correct at high temperature and is still used in modern texts.

worked through crudely, leading to results that he found puzzling. When Schrödinger wrote back to Einstein on November 3, in addition to applauding Einstein's development of Bose statistics he proposed to carry through Einstein's alternative approach in detail, which he was able to do in a scant few days. He was less troubled than Einstein by the answer he found, which confirmed Einstein's original argument, and proposed a joint publication: "the basic idea is yours . . . and you must decide about the further fate of your child. . . . I need not emphasize the fact that it would be a great honor for me to be allowed to publish a joint paper with you."

A touching exchange ensued, in which Einstein insists he should not be a coauthor, "since you have performed the whole work; I feel like an 'exploiter,' as the socialists call it." Schrödinger immediately demurs: "not even jokingly would I have . . . thought of you as an "exploiter" . . . one might say: 'when kings go building, wagoners have work.'" On December 4 he sent Einstein a complete draft of the paper with the second author slot blank, but Einstein presented the paper to the Prussian Academy without his own signature. Einstein's intuition was again right; there were subtle errors in the reasoning, and the approach itself turned out to be unwieldy, in contrast to Einstein's first approach, which is still found in modern texts.

However, only a couple of weeks later, Schrödinger had yet another ideal gas paper in the works, which led directly to the wave equation. Just prior to his letter to Einstein in early November, Schrödinger had finally managed to get hold of de Broglie's thesis, which he had sought because of Einstein's strong endorsement of it in his work on the quantum gas. Having now got hold of the thesis, Schrödinger tells Einstein that "because of it, section 8 [the wave-particle section] of your second [quantum gas] . . . paper has also become completely clear to me for the first time." Two weeks later, writing to another physicist, he remarks that he is very much inclined toward a "return to wave theory."

His final ideal gas paper, submitted December 15, just before he left for his Christmas holiday, was a last effort to digest Einstein's new quantum statistics; it is titled simply "On Einstein's Gas Theory." He sets out to derive the same answers as Einstein without accepting the strange state counting of Bose, which he thinks requires too great a "sacrificium intellectus." The way out, he says, is "nothing else but taking seriously the De Broglie–Einstein wave theory of the moving particles, according to which the particles are nothing more than a kind of 'white crest' on a background of wave radiation." He then formulates the mathematical problem differently from Einstein but shows how to reach exactly the same final equations. From the perspective of a modern physicist[10] the two calculations are equivalent and have the same meaning, but Schrödinger felt that by interpreting the fundamental objects as waves instead of as particles the weirdness of indistinguishability was somehow made more palatable. He concludes by speaking of the particles as "signals" or "singularities" embedded in the wave, highly reminiscent of Einstein's failed ideas of 1909–1910 and his later idea of "ghost fields" guiding the particles developed in the early 1920s. Except that Schrödinger clearly now thinks that the particles are the "ghosts," the ephemera, since by viewing waves as fundamental he has "explained" Bose-Einstein statistics in a natural way. Later he would say, "wave mechanics was born in statistics."

A few days before Christmas 1925 Schrödinger set off to a familiar mountain lodge in the Swiss village of Arosa, determined to find a new equation to describe these matter waves. Although he took his skis (he was an expert alpinist) and was accompanied by an unnamed "old girlfriend,"[11] it seems that this trip was really focused on wave equations. De Broglie, despite Einstein's praise, had not produced a new governing equation, similar to Maxwell's electromagnetic wave equation, that could predict or explain the remaining mysteries of the atom. He had produced some suggestive mathematical relations, which made contact with the old quantum theory of Bohr and Sommerfeld, but only at the most elementary level. The central mystery of quantum

---

[10] Einstein immediately saw this equivalence, writing to Schrödinger after his paper appeared, "I see no basic difference between your work on the theory of the ideal gas and my own."

[11] Hermann Weyl, a distinguished physicist and a friend of Schrödinger's, later famously commented that Schrödinger "did his great work during a late erotic outburst in his life."

theory, *quantization*, was not really resolved by de Broglie's work. Why wasn't nature continuous? Why are only certain energies allowed for electrons bound to atomic nuclei?

Schrödinger saw an answer. Classical waves, or vibrations, in a confined medium have certain natural constraints on their properties. Consider a violin string of a certain length, L, clamped down at each end. The notes it can play arise from vibrations of the strings; these vibrations are waves of sideways displacement in the string, but they cannot have a continuously varying wavelength as is possible for waves in an open (essentially infinite) medium. The longest wavelength, λ, they can have is twice the string length. Why is this the longest? First, consider that, for any wave on a string, the string's displacement must be zero at the points where it is clamped down. The simplest form of displacement the string can have away from these fixed points is for it to be displaced everywhere in the same sideways direction (left or right) at a given instant, so that the maximum displacement is in the middle and it decreases back to zero at both clamped ends. This displacement then oscillates back and forth, causing sound waves of a certain pitch. Since we measure wavelength by the distance between points in the medium that take us through a peak *and* a trough, this shape corresponds to half the full wavelength, so the wavelength is $\lambda = 2L$. This will determine the lowest note the violin can play (for a given string tension). The next-lowest note will have $\lambda = L$, implying that there will be no displacement at the center of the string (even though that point isn't clamped down). In general the only allowed wavelengths are $\lambda = 2L/n$, where n is a whole number ($n = 1, 2, 3 \ldots$). That's the point: for confined waves, nature produces whole numbers automatically. De Broglie had hinted at this, but now Schrödinger realized that this was the key to getting the quantum into quantum theory.

The mathematics of the old quantum theory had not done this in a natural way. Bohr and his followers had taken mathematical expressions that are continuous (i.e., don't involve whole numbers exclusively) and had simply restricted them to whole numbers by fiat. Schrödinger, by contrast, was looking for an equation that simply *did not have* continuous solutions, one in which each solution would be connected to a whole number organically.

As noted above, wave equations, through the quantization of the wavelength, have this property. They are differential equations, describing continuous change in space and time; but when the waves are confined, only certain wavelengths are physically possible. So Schrödinger was looking for a matter-wave differential equation to describe electrons: it would have to describe a matter field and an extended "disturbance" of the field that varied in space and time but was confined to the vicinity of the atomic nucleus by the electron's attraction to the nucleus.

Maxwell's electromagnetic wave equation was the only fundamental wave equation of physics at that time, but because photons are massless, it had no safe place for the introduction of Planck's constant (as Einstein had learned to his dismay fifteen years earlier). Schrödinger's challenge was to fashion a wave equation, modeled on Maxwell's, that included Planck's constant as well as the physical constants e and m, representing the charge and mass of the electron. Maxwell's wave equation does, however, contain the wavelength, λ, of the EM waves which then lead to quantized values and whole numbers, when it is written for "standing waves" confined to a specific region. So the key idea was to write an equation similar to the electromagnetic wave equation but for a new wave field, a "matter wave" described by a mathematical expression now known as the "wavefunction," and to replace λ using de Broglie's relation $\lambda = h/p = h/mv$.

With this approach Schrödinger had an equation containing h, Planck's constant, and m, the electron mass. The electron velocity, v, can be eliminated from the equation in favor of the difference between its total energy and potential energy. The potential energy of the electron orbiting the nucleus of course depends on its charge, e. The resulting "time-independent Schrödinger equation" for hydrogen contains the "holy trinity," h, m, and e. And now the moment of triumph: since only certain wavelengths are allowed, that implies that the only unknown in the equation, the total energy of the electron, can only take on certain allowed values. The energy of electrons in an atom *is* quantized, not by fiat, but due to the fundamental properties of waves confined to a fixed region in space.[12]

After returning from his "ski" trip, Schrödinger immediately subjected his new equation to the acid test: could it reproduce the energy levels, grouped into different "shells," for the case of atomic hydrogen, which were known from spectral measurements for decades and "explained" in an ad hoc manner by the old quantum theory? The answer was a resounding yes. The details of the calculations took only a few weeks, and on January 27, 1926, the first of Schrödinger's seminal papers was received at the *Annalen der Physik*. It states the breakthrough thus: "in this paper, I wish to consider . . . the simple case of the hydrogen atom . . . and show that the customary quantum conditions can be replaced by another postulate, in which the notion of 'whole numbers' . . . is not introduced. . . . The new conception is capable of generalization and strikes, I believe, very deeply at the true nature of the quantum rules."

After presenting the detailed solution of his hydrogen equation, he briefly touches on its interpretation and its origins. "It is, of course, strongly suggested that we try to connect the [wavefunction] with some vibration process in the atom, which would more nearly approach reality than the electron orbits [of Bohr-Sommerfeld theory]," but he feels that this is premature, since the theory needs further development. However, "Above all, I wish to mention that I was led to these deliberations by the suggestive papers of M. Louis de Broglie . . . I have lately shown that the Einstein gas theory can be based on [such] considerations . . . the above reflections on the atom could have been represented as a generalization from those on the gas model." Later Schrödinger would say, "My theory was inspired by L. de Broglie . . . and by brief, yet infinitely far-seeing remarks of A. Einstein."

This first paper was followed in rapid succession by five more in just six months, in which Schrödinger, the consummate craftsman,

---

[12] This simple argument, while certainly part of Schrödinger's reasoning, was not how he first presented the equation and omits his failed initial attempt to come up with an equation consistent with Einstein's relativity theory.

working alone, determined essentially all the known properties of atomic spectra from the solutions of his wave equation. It was a breathtaking display, about which even his competitor Born would later remark, "what is more magnificent in theoretical physics than Schrödinger's six papers on wave mechanics?" The normally reserved Sommerfeld called Schrödinger's equation "the most astonishing among all the astonishing discoveries of the twentieth century." Hence, by June of 1926, physicists had uncovered most of the new laws and mathematical methods necessary to describe physics on the atomic scale; they just didn't yet know what they meant. However, an interpretation was soon to emerge, one that would challenge the philosophical principles that both Schrödinger and Einstein held dear.

CHAPTER 28

CONFUSION AND THEN UNCERTAINTY

If we are still going to have to put up with these damn quantum jumps, I am sorry that I ever had anything to do with quantum theory.
—SCHRÖDINGER TO BOHR, OCTOBER 1926

"I am convinced that you have made a decisive advance with your formulation of the quantum condition, just as I am equally convinced that the Heisenberg-Born route is off the track." Thus Einstein wrote to Schrödinger in late April of 1926. The Heisenberg-Born route, a different approach to the "quantum conditions," introduced the term "quantum mechanics" as a more rigorous replacement for the nebulous conceptual structure of "quantum theory." This method had begun to bear fruit six months earlier than Schrödinger's, and unlike his work it arose independently of Einstein's recent successes with the quantum gas.

It represented the radical point of view that since atoms were, practically speaking, impossible to observe in space and time, one should stop attempting to describe them by space-time orbits as in classical mechanics. Instead one should develop a description in terms of observable atomic variables, which might not themselves be easily visualized, such as the absorption frequencies for light incident on the atom, and how strongly each frequency was absorbed. The first breakthrough using this approach had come from the twenty-three-year-old prodigy Werner Heisenberg, who formulated his method in July of

1925. By the time of Schrödinger's work, Einstein had been ambivalently struggling with this new framework for quite some time already, since Heisenberg was working in the research group of his close friend Max Born in Göttingen. Born had immediately informed him of Heisenberg's initial sighting of a New World of the atom, writing in a letter dated July 15, 1925, that Heisenberg's paper "appears rather mystifying, but is certainly true and profound."

Heisenberg was not the only young genius to find his way to Born's research team in Göttingen. Six of Born's research assistants and one of his PhD students would go on to win the Nobel Prize,[1] and three of them—Enrico Fermi, Wolfgang Pauli, and Heisenberg—would contribute cornerstones to the rising quantum edifice. Born, only three years younger than Einstein, was from Prussian Silesia, and was of Jewish descent (like many of Einstein's closest friends). He had been appointed associate professor at Berlin from 1915 to 1919, arriving just in time to observe Einstein's awe-inspiring success with general relativity theory. He, and his wife Hedwig, formed a lifelong friendship with Einstein, although Born maintained as well a certain reverence for his friend, whom he would refer to, after his death, as "my beloved master." Born made seminal contributions to physics and eventually won the Nobel Prize himself, but he was not an imposing intellect, and he sometimes had trouble keeping up with his brilliant wards. Of Pauli, who was renowned for his critical brilliance, he said, "I was from the beginning quite crushed by him . . . he would never do what I told him to do." Heisenberg, he recalled, was quite different: "he looked like a simple peasant boy, with short, fair hair, clear bright eyes and a charming expression" when he arrived, "very quiet and friendly and shy. . . . Very soon I discovered he was just as good in the brains as the other one."

After a few months spent visiting Niels Bohr in the fall of 1924, Heisenberg returned to Göttingen with the germ of an idea for a

---

[1] In addition to Heisenberg, Pauli, and Fermi, the others were Max Delbruck (PhD), Eugene Wigner, Gerhard Herzberg, and Maria Goeppert-Mayer, the second woman to win the prize in physics.

FIGURE 28.1. Werner Heisenberg circa 1927. AIP Emilio Segrè Visual Archives, Segrè Collection.

completely new quantum theory of the atom, distinct from the old Bohr-Sommerfeld approach. This approach, while it worked for hydrogen and a few other atoms, appeared to be breaking down for more complicated atoms and molecules. In fact, by 1924 more than a decade had passed since Bohr's pathbreaking work, and a full quarter century since that of Planck; many physicists were beginning to wonder if the fundamental laws of the atom were simply beyond human ken. In May of 1925 the enfant terrible, Pauli, wrote despairingly to a friend, "right now physics is very confused once again—at any rate it's much too difficult for me and I wish I were a movie comedian or some such." However, Heisenberg was just about to shake the field out of its malaise.

Heisenberg's idea was to take the continuous trajectory of a particle, which in classical physics is represented by the three Cartesian coordinates x, y, z that vary continuously with time, and replace each coordinate with a list of numbers arrayed in rows and columns, rather like a Sudoku puzzle. Each number in the list is not fixed, but oscillates in time sinusoidally, with a characteristic frequency. When applied to electrons in an atom, the frequencies corresponded to the observable "transition frequencies" at which the atom would absorb and emit light. First, however, Heisenberg considered the most basic "toy problem" of mechanics, the familiar linear harmonic oscillator (mass on a spring). He was able to show that using his new definition for position, and a similar one for momentum, the energy of the oscillator was conserved; that is, it didn't change in time as long as the energy took the special values found by Planck so long ago, quantized in steps of hv (where v is the frequency of the oscillator). So here, in Heisenberg's new arithmetic, the whole numbers of quantum theory also arose naturally from the math and were not imposed externally, just as they would later appear naturally in Schrödinger's wave approach. Heisenberg first discovered this while recovering from an allergy attack on the North Sea island of Helgoland, and he was so excited that he stayed up all night working, and then, lying on a rock watching the sun rise, he thought to himself, "well something has happened."

Indeed something had. This mode of thinking was simply orthogonal to everything physicists had been trying to do in atomic theory, and it broke the impasse. Heisenberg informed his friend Pauli, who was elated, saying that the idea gave him renewed "*joie de vivre* and hope . . . it's possible to move forward again." Heisenberg wrote up his initial ideas with the boldly stated goal of establishing a new quantum mechanics, "based exclusively on relationships between quantities which are in principle observable." Born, with another talented student, Pascual Jordan, quickly realized that Heisenberg's "lists of numbers" were objects that mathematicians refer to as matrices, and that the rule for combining them that Heisenberg had invented was the known rule for multiplying matrices. An odd thing about representing physical magnitudes by matrices is that when multiplying matrices, in general x times y is not equal to y times x. This curiosity would end up having a deep significance in the final theory. Within a few months Born, Heisenberg, and Jordan were able to put together a definitive paper announcing the rules for calculating observable quantities in the new quantum mechanics, which, in their version, would become known also as "matrix mechanics."

Einstein, despite Born's endorsement, reacted suspiciously to the breakthrough from the beginning, writing to Ehrenfest in September 1925 with a typically earthy judgment: "Heisenberg has laid a big quantum egg. In Göttingen they believe in it (I don't)." Despite his skepticism, he realized that a substantial advance had been made, telling Besso in December 1925 that matrix mechanics was "the most interesting thing that theory had produced in recent times"; but he could not resist a dig at its odd structure, "a veritable witches' multiplication table . . . exceedingly clever and because of its great complexity safe against refutation." Sarcasm notwithstanding, he studied the theory closely and discovered several technical objections, which he communicated to Jordan.[2]

Bose recalled that upon his arrival in Berlin in the fall of 1925, "Heisenberg's paper came out. Einstein was very excited about the new quantum mechanics. He wanted me to try to see what the statistics of light-quanta . . . would look like in the new theory." But Einstein's reservations were beginning to win out; early in 1926 he wrote to Ehrenfest, "more and more I tend to the opinion that the idea, in spite of all the admiration [I have] for [matrix mechanics], is probably wrong." Just as he was hardening his negative view, in January the newly reenergized Pauli showed how to derive the basic hydrogen spectrum using matrix mechanics, an apparently decisive proof that the theory was on the right track. Of course Schrödinger was just at that time deriving the *same result* by the quite different method of his wave equation.

Schrödinger's approach was superficially much more congenial to the classical physics worldview, based as it was on a continuum wave equation in space and time, similar to that of Maxwell, and seeming to arrive at quantized energies via the familiar properties of vibrating waves. Einstein, Planck, Nernst, and Wien, the reigning royalty of German physics, all jumped on the Schrödinger bandwagon immediately. Born, now a bit under siege, later recalled that Schrödinger's paper "made much more of an impression than ours. It was as though ours didn't exist at all. All the people said *now* we have the *real* quantum mechanics." However, Born would soon have a key ally; Niels Bohr had been moving toward a view that the conventional space-time picture of the atom was fatally flawed, and his force of personality would eventually prevail, although not without some further twists and turns.

---

[2] Einstein wrote a number of letters to Heisenberg in this period, all of which have been lost. At least one, Heisenberg recalled, was signed "in genuine admiration."

Initially the two sides believed that they were faced with a choice between two fundamentally different theories, so that Einstein, in the same letter to Ehrenfest in which he called matrix mechanics "probably wrong," described Schrödinger's innovation as "not such an infernal machine [as matrix mechanics], but a clear idea—and logical in its application." And a few weeks later, in early May, he told Besso, "Schrödinger has come out with two excellent papers on the quantum rules, which present some profound truths." But the period of either/or decision making was brief. A dramatic change in the debate occurred at one of the famed Berlin colloquia, where Einstein often presided. A young student, Hartmut Kallmann, recorded the events. "People were packed into the room as lectures on Heisenberg's and Schrödinger's theories were given. At the end of these reports Einstein stood up and said, 'Now just listen! Up until now we have had no exact quantum theory, and now suddenly we have two. You will agree with me that these two exclude each other. Which theory is correct? Perhaps neither is correct.' At that moment—I shall never forget it—Walter Gordon stood up and said: 'I have just returned from Zurich. Pauli had proved that the theories are identical.'"[3] Actually by mid-March Schrödinger, prior to Pauli (who never even bothered to publish his proof), was able to show that the equations of matrix mechanics followed from his wave equation and vice versa; matrix mechanics could be used to derive the Schrödinger equation. The two theories were indeed mathematically equivalent.

At this point the debate shifted to the question of the *meaning* of the new theory, and the aesthetic and conceptual merits of the two different formulations. Already, in his paper proving their equivalence, Schrödinger had slipped in a jibe against the matrix approach, saying that he was "discouraged, if not repelled" by the difficulty of its methods and its lack of transparency. And he repeatedly stated that his approach was the more "visualizable," prompting a fed-up Heisenberg to declare in a letter to Pauli, "what [he] writes about Anschaulichkeit [visualizability] makes scarcely any sense. . . . I think it is crap."

---

[3] Kallmann most likely erred in his memory of the city, since Gordon was working with Pauli in Hamburg at that time. Schrödinger was in Zurich.

Matters came to a head in July, when Schrödinger made a "victory tour" of the conservative physics centers of Berlin, where they had begun recruiting him to replace Planck, and Munich, where Wien and Sommerfeld were in charge. By coincidence Heisenberg was in Munich when Schrödinger spoke, and he raised some unresolved issues for wave mechanics in the question period at the end of the lecture. Before Schrödinger could respond, Heisenberg was almost "thrown out of the room" by Wien, who thundered, "young man, Professor Schrödinger will certainly take care of all these questions in due time. You must understand that we are now finished with all that nonsense about quantum jumps." A shaken Heisenberg wrote immediately to Bohr, who responded by inviting Schrödinger to Copenhagen. A marathon session of conceptual arm wrestling ensued, ending with Schrödinger in bed exhausted and sick, but unconverted. The key point that Bohr insisted upon is that while Schrödinger's wave equation appeared to restore a continuous description of nature, when applied to atoms it would inevitably lead back to the fundamental discontinuity of natural processes implied by quantum phenomena. At about the same time Einstein and his close friend Max Born were wrestling with exactly this issue.

For Einstein, the mathematical equivalence of the two theories simply extended his doubts about matrix mechanics to wave mechanics. He was not immune to the exhilaration felt by his colleagues as the historic puzzles of atomic structure were being unraveled almost on a weekly basis. After another colloquium, at which the evidence for the newly discovered spin of the electron was presented, Bose ran into him on a streetcar: "we suddenly found him jumping [into] the same compartment where we were, and forthwith he began talking excitedly about the things we have just heard. He has to admit that it seems a tremendous thing, considering the lot of things which these new theories correlate and explain, but he is very much troubled by the unreasonableness of it all. We were silent, but he talked almost all the time; unconscious of the interest and wonder that he was exciting in the minds of the other passengers."

The unreasonableness that Einstein felt now focused mainly on the meaning of Schrödinger's wavefunction, which somehow represented the behavior of electrons bound to atomic nuclei. Schrödinger originally tried to argue that his matter waves could accumulate in a localized region of atomic dimensions, carrying along a bump or "crest" that behaved like a particle. But further study soon showed that such a "wave packet" could not cohere over long times; the math was actually very similar to that of light waves, and the failure of this idea reprised Einstein's own failure to find particulate behavior in Maxwell's wave equation back in 1910. It is likely that Einstein spotted this problem very quickly. A fallback position, taken by Schrödinger subsequently, was to assert that there simply *are* no electron particles; the "real electron" is a wave of electric charge density, spread out in space on dimensions somewhat larger than the atom. But there was a further basic problem with this picture. Einstein expressed this in June of 1926 in a letter to a colleague, Paul Epstein: "We are all here fascinated by Schrödinger's new theory of quantum levels . . . strange as it is to introduce a field in q-space, the usefulness of the idea is quite astonishing." What was this "q-space" that Einstein found so strange?

All the waves that were known to physicists at that time were represented as an oscillating field or disturbance in our normal three-dimensional space, even electromagnetic waves, which, according to Einstein, didn't require a medium (the ether) to exist. The number of particles in a wave did not enter the equation except through the density of the medium, as for sound waves, or through the intensity of the wave, for electromagnetic waves. But electron waves could not be represented that way. Isolated free electrons could be studied, and their charge could be measured; it was $-e$, the same in magnitude (and opposite in sign) from that of the proton. The periodic table of elements requires that hydrogen have a single electron, helium two electrons, lithium three, and so on. So to describe electrons in helium, for example, one needed to have a wave equation for *two* electrons in different quantum states, with total electric charge adding up to 2 times $-e$. There was only one way to do this, mathematically speaking, using Schrödinger's equation: the "two-electron" wavefunction had to "live" in a six-dimensional space, three dimensions for the first electron and three more for the second. Moreover, the wavefunction for electrons in a large atom such as uranium-235 had to live in a 705-dimensional space! This was "q-space," an abstract space that "copies" our three-dimensional space N times in order to represent N electrons. Einstein recognized this strange feature in his very first letter to Schrödinger, and to him it was an enormous clue that a classical wave picture might not be restored through Schrödinger's equation.

Einstein was not the only person puzzling over how to interpret Schrödinger's waves. Max Born also had grave reservations about the idea that the electron was a conventional wave. He worked closely with the noted experimentalist James Franck, who did measurements of electron beams colliding with atoms. "Every day," he recalled, "I saw Franck counting particles and not measuring continuous wave distributions." The moment he learned about the Schrödinger equation, he had an intuition about what its matter waves represented. He recalled Einstein's idea that the electromagnetic field is a "ghost field" guiding the photons. "I discussed this with him very often. He said that as long as there was nothing better, one can [use this approach]." But now, for matter waves, Born felt this *was* the true picture: the Schrödinger wavefunction represented a guiding wave of *probability*. Mathematicians and physicists were already used to the idea of assigning probabilities to a continuous space, essentially by dividing the space into infinitesimal regions. Born argued that Schrödinger's wavefunction represented such a probability density,[4] which actually moved deterministically in space as a wave but simply described how *likely* it would be to find an electron particle in each particular region of space.

By the end of June 1925 Born went public with his idea, submitting a paper titled "Quantum Mechanics of Collision Phenomena." In it he formulates the problem of a directed matter wave (representing a stream of electrons in a beam directed at an atom) that interacts with the electric field of the atom and then "scatters" in all directions, just as water waves hitting a post send out circular waves in all directions. Could this really mean that each electron "breaks up" and goes in all

---

[4] More precisely, it is the absolute square of the wavefunction that represents a probability density.

directions like a smeared-out electrical "oil slick"? That is exactly the view that Schrödinger wants to take, but Born is having none of it. He insists that the expanding circular wave just determines the probability of finding a whole, pointlike electron emerging in particular direction. To test this idea you need to do the same experiment over and over again and count the number of electrons that go in each direction. "Here the whole problem of determinism arises. From the point of view of our quantum mechanics, there exists no quantity which *in an individual case* causally determines the effect of the collision. . . . I myself tend to give up determinism in the atomic world." In his view we need to adopt a weaker form of determinism: "the motion of particles follows probabilistic laws, but the probability itself propagates according to the law of causality."

When Schrödinger learned of Born's interpretation, he was incensed and engaged him in an "acrimonious debate." As Born recalled, "he believed that [matter waves] meant some continuous distribution of matter and I was very much opposed to it [because of Franck's experiments]. . . . he was very offensive, as he always was when somebody objected to [his ideas]." Schrödinger's opposition notwithstanding, the Born probabilistic interpretation of the wave function was widely adopted almost immediately, and was the basis of Born's eventual Nobel Prize. However, the person who had inspired Born's critical step, Einstein, was among the few holdouts. In November of 1926 Born wrote to his dear friend: "I am entirely satisfied, since my idea to look upon the Schrödinger wave field as a '[ghost field]' in your sense proves better all the time. . . . Schrödinger's achievement reduces itself to something purely mathematical; his physics is quite wretched." But by this time Einstein's reservations had solidified into an unshakable conviction. Just a few days later he sent Born his famous and crushing response: "Quantum mechanics calls for a great deal of respect. But some inner voice tells me that this is not the true Jacob. The theory offers a lot, but it hardly brings us closer to the Old Man's secret. For my part, at least, I am convinced he doesn't throw dice."

Some months earlier Einstein had met privately with Heisenberg to discuss quantum mechanics. Heisenberg had presented his view that the new theory should restrict itself to describing observable quantities, and not unobservable electron orbits. Einstein rejected this view, leading Heisenberg to rejoin, "isn't that precisely what you have done with relativity theory." Einstein responded, "possibly I did use this form of reasoning . . . but it is nonsense all the same.[5] . . . It is the theory which decides what can be observed." This conversation stuck with Heisenberg, and a year later, while pondering the meaning of quantum mechanics, it came back to him. "It must have been one evening after midnight when I suddenly remembered my conversation with Einstein, and particularly his statement, 'it is the theory which decides what we can observe.' I was immediately convinced that the key to the gate that had been closed so long must be sought right here." Within days he had used the new quantum mechanics to prove his uncertainty principle. One could observe the position of an electron very accurately, or the momentum of an electron very accurately, but not both at the same time. That's what the theory had decided. Even this realization, so fiercely opposed by Einstein, had been stimulated by his own insight.

---

[5] Later, in replying to the same reproach from his friend Philipp Frank, Einstein responded with the pithy retort, "A good joke should not be repeated too often."

CHAPTER 29

NICHT DIESE TÖNE

All the fifty years of conscious brooding have brought me no closer to the answer to the question, "what are light quanta?" Of course today every rascal thinks he knows the answer, but he is deluding himself.
—EINSTEIN TO BESSO, 1951

"Here I sit in order to write, at the age of 67, something like my own obituary . . . [this] does . . . not come easy—today's person of 67 is by no means the same as was the one of 50, of 30 or of 20. Every reminiscence is colored by today's being what it is, and therefore by a deceptive point of view." Einstein, in the autobiographical sketch he thus begins, confirms his initial disclaimer. Readers hoping to learn from the man himself amusing anecdotes or details of his personal life were disappointed; the article of forty-six pages is a rather dense treatment of his philosophy of science, the evolution of physical theory, and then his actual contributions to science, ending with a technical statement of his latest attempt at a unified field theory. However, his revolutionary work on light quanta, and his groundbreaking quantum theory of specific heat of 1905–1907, merit only one long sentence. His early discovery of wave-particle duality gets a bit less than one page, ending in a remark that the current quantum theoretical explanation for it is "only a temporary way out." His foundational work on the quantum theory of radiation and the spectacular discovery of Bose-Einstein condensation get no mention at all. He devotes much more space to his critique of quantum mechanics than to his contributions thereto. In contrast, relativity theory, special and general, is laid out in beautiful and exacting detail.

After the decisive year of 1926, in which he rejected the new quantum theory as the ultimate description of reality, he briefly sought to show, via his classic method of gedankenexperiments, that the theory contained internal contradictions. However, fairly soon he accepted the consistency of its logical structure with the comment "I know this business is free of contradictions, but in my view it contains a certain unreasonableness." By September of 1931 he would graciously nominate both Heisenberg and Schrödinger for the Nobel Prize, with the comment "I am convinced that this theory undoubtedly contains a part of the ultimate truth."

But despite this grudging endorsement, Einstein himself never applied the quantum formalism to a specific physics problem for the rest of his career, except in the context of a famous critical paper written in 1935 with younger collaborators, Podolsky and Rosen. The article drew attention through a thought experiment to the "spooky action at a distance" implied by quantum theory, which the authors claimed made the theory an incomplete description of reality. Modern realizations of this "EPR" experiment have fully confirmed the existence of this effect, a counter-intuitive correlation between distant particles. Such effects are referred to as "entanglement"; they form the basis of much of the new field of quantum information science. Many now consider Einstein's recognition and prediction of the EPR effect as his last major contribution to physics.

Not only Einstein but also de Broglie and Schrödinger, the two quantum pioneers whom he had championed, made little contribution to the further application of quantum theory, and both ended up joining Einstein in rejecting it on philosophical grounds. As a consequence, the history of the discovery/invention of quantum theory was told from the perspective of Bohr, Heisenberg, Born, and their legions of students and collaborators. (Einstein, de Broglie, and Schrödinger *had* no students or collaborators in their works on quantum theory.) The matrix mechanicians, whose approach was instantly devalued following Schrödinger's discovery of the "real quantum mechanics," simply appropriated that work and gave it the interpretation that fit their understanding (and, it must be admitted, the experimental evidence). Ironically, Schrödinger was correct; his method *was* much more intuitive and visualizable than that of Heisenberg and Born, and it has become the overwhelmingly preferred method for presenting the subject. But with Born's probabilistic interpretation of the wavefunction, Heisenberg's uncertainty principle, and Bohr's mysterious complementarity principle,[1] the "Copenhagen interpretation" reigned supreme, and the term "wave mechanics" disappeared; it was all quantum mechanics. The limitations on human knowledge of the physical world implied by these concepts were accepted by all practicing physicists. To this new generation Einstein became known primarily for relativity theory, admired by all, and secondarily for his stubborn refusal to accept the elegant new atomic theory of everything.

However, if one takes stock of the conceptual pillars of the new theory, in light of the historical record, a rather different picture emerges. Einstein surely shares with Planck the discovery of quantization of energy, as Planck never accepted that the quantum of action implied quantization of mechanical energy until many years after Einstein had become the first to proclaim it. It was Einstein who first realized that quantized energy levels explained the specific heat of solids, which justified the Third Law of thermodynamics and brought chemists such as Nernst into the quantum arena. Einstein, in his paper on light quanta, discovered the first force-carrying particles, photons, now the paradigm for all the fundamental forces. Following up on this, he discovered the wave-particle duality of light and, in 1909, based on his rigorously correct fluctuation argument, predicted that a "fusion theory" must emerge to reconcile the two views. In 1916 his quantum theory of radiation combined the ideas of Bohr, Planck, and his own light quanta to put Planck's blackbody law on a firm basis. Here he introduced, for the first time, the core concept of intrinsic randomness in atomic processes, which the mature

---

[1] A philosophical principle about the impossibility of a unified picture of the atomic world, the utility of which is controversial

theory would accept as fundamental. He also introduced the notion of the *probability* to make a quantum jump, and he distinguished between spontaneous and stimulated transitions, ideas fundamental to, for example, the invention of the laser. And during 1924–1925 he elevated Bose statistics from obscurity, explained what it meant and why it had to be correct, and derived the mind-boggling condensation phenomenon it implied, something undreamt of by Bose himself. Finally, without ever publishing it, he developed the rule of thumb that electromagnetic wave intensity could be thought of as determining a probability to find photons in a certain region of space, the idea that stimulated Born's crucial interpretation of matter waves.

In summary: quantization of energy, force-carrying particles (photons), wave-particle duality, intrinsic randomness in physical processes, indistinguishability of quantum particles, wave fields as probability densities—these are most of the key concepts of quantum mechanics. As Born would later say, "Einstein is therefore clearly involved in the foundation of wave mechanics and no alibi can disprove it." The magnitude of these achievements? Four Nobel Prizes would be about right, instead of the one he received, grudgingly, in 1922. Not that Einstein cared much for such accolades.

Why did Einstein, who clearly understood the structure of the new theory and the necessity of introducing radical concepts to explain the atom, refuse to accept that theory and hold out for a very different resolution of the quantum dilemma? In my opinion this was the result of both his life experiences in doing science and his fundamental motivation for choosing that life.

Twice in his scientific career Einstein had wandered so far from the mainstream that even the many colleagues who already regarded him as an historic genius simply dismissed his views as wildly speculative and not to be taken seriously. Special relativity was not such a case, building as it did on the work of Lorentz and others, although certainly it unveiled a spectacular physical and epistemological insight that was uniquely Einsteinian. The first time was, of course, the light quantum "nonsense," for which Planck felt compelled to apologize when nominating him to the Prussian Academy, and which Bohr still ridiculed a full two decades after it was proposed. The second time was with the theory of general relativity. In the latter case the idea did not elicit ridicule but simply incomprehension. There was no crisis in gravitation theory that required a radical resolution; what was this eccentric mastermind doing anyway?

The development of general relativity had proceeded unevenly, with dead ends and backtracking, technical errors ultimately corrected, and then an epiphany as the beautiful final equations emerged and predicted correctly the precession of Mercury and the bending of starlight. Einstein recalled the struggle thus: "the years of searching in the dark for a truth that one feels but cannot express, the intense desire and the alternation of confidence and misgiving until one breaks through to clarity and understanding, are known only to him who has experienced them himself." Much later, in rejecting the modern quantum theory, he remarked, "it is my experiences with the theory of gravitation which determines my expectations."

Moreover, just before the promulgation of the new quantum theory, in the summer of 1925, Einstein had experienced a similar vindication of his faith in the existence of light quanta. In 1924 Bohr and collaborators had put forth a new approach to the interaction of light and matter that sacrificed the principle of conservation of energy and momentum and introduced statistical considerations into the theory, although not in a manner that would turn out to agree with the final quantum theory. In print Bohr flatly stated that the theory of light quanta was "obviously not [a] satisfactory solution of the problem of light propagation." Einstein staunchly opposed Bohr's new theory, since he believed that the conservation laws must be exact or his beloved thermodynamics would be undermined. He also pointed to dramatic recent experiments by the American physicist Arthur Holly Compton that seemed to confirm the conservation laws for the collisions of an x-ray photon with an electron (treated as particles). However, it still could be argued that Compton's experiments left open the possibility that momentum and energy were only conserved *on average* and not in each individual collision. This possibility was ruled out late in 1924 by landmark experiments of Bothe and Geiger in which the individual collisions were measured and shown conclusively to obey the conservation laws for two particles.

In January of 1925 Born wrote to Bohr, "the other day I was in Berlin. There everybody is talking about the result of the Bothe-Geiger experiment, which decided in favor of light-quanta. Einstein was exultant." In April of 1925, two months before Heisenberg's *coup de destin*, Bohr conceded that it was time "to give our revolutionary efforts [to banish light quanta] as honorable a funeral as possible." Einstein's apparently infallible intuition had triumphed one last time. Thus when the newest statistical theory of the atom, the Heisenberg-Born-Schrödinger synthesis, commandeered the stage, Einstein must have felt a sense of déjà vu. Just hold out long enough, and again he would be proved right.

Einstein's most famous objection to the theory was the "dice complaint": its insistence on the intrinsic randomness of individual events and the abandonment of rigid causality. But Schrödinger's q-space picture had actually undermined Einstein's objection to a probabilistic theory. Eugene Wigner, a leading figure in the second generation of quantum pioneers, was studying in Berlin in the early 1920s, and recalled that Einstein was quite "fond" of his guiding field concept, "[which] has a great similarity with the present picture of quantum mechanics," but "he never published it . . . [because] it is in conflict with the conservation principles." However, in the N-dimensional space of Schrödinger's waves, the conservation laws *survived*, even if the outcomes were indeterminate. In quantum mechanics, if two particles collide, even if one has full knowledge of the particle properties before the collision, it is impossible to predict, in each individual case, in which directions the two particles will be traveling after the collision. One can only state the *probability* that they emerge from the collision in a certain pair of directions. Nonetheless, there is *zero* probability that the two particles emerge with a different total momentum and energy than they went in with. They are like a magic pair of coins, which when flipped individually give you heads or tails randomly and with equal probability, but when flipped as a pair always come up with opposite faces showing. So quantum indeterminacy still respects the conservation laws.

Perhaps for this reason Einstein's later critiques of quantum theory focused less on its indeterminacy and more on its strange epistemological status. In quantum mechanics the actual act of measurement is part of the theory; those magic coins just mentioned exist in a state of (heads, tails)-(tails, heads) uncertainty until they are measured, and then they are forced to "decide" which state they are in. This is true even if the coins are flipped very far apart, implying that obtaining knowledge of one coin, through measurement, "changes" the state of the other coin an arbitrary distance away. This is the "spooky action at a distance" that Einstein detested, now known as "quantum entanglement." But beyond its apparent tension with relativity theory, the entire conceptual structure seems to break down the barrier between the "real world" of objective nature and the subjective world of human perception. "Do you really believe that the moon exists only if I look at it?" he used to say. Such a notion fundamentally challenged Einstein's credo.

In his autobiography he states, "Physics is an attempt to conceptually grasp reality as it is . . . , independently of its being observed." In a letter to Born, late in his life, he amplified on this theme: "We have become Antipodean in our scientific expectations. You believe in the God who plays dice and I in complete law and order in a world which objectively exists, and which I, in a wildly speculative way, am trying to capture." The importance of this dichotomy, the transitory, subjective, and ultimately insignificant individual versus the eternal order of the Cosmos, was central to his personal philosophy. As a very young man he rejected the "nothingness of the hopes and strivings which chases most men restlessly through life." In this way one could "satisfy the stomach" but not the "thinking and feeling being." But he soon realized that there was another way to live: "out yonder there was this huge world, which stands independently of us human beings and which stands before us like a great eternal riddle, at least partially accessible to our thinking and inspection. The contemplation of this world beckoned us like a liberation." When he had risen to the apex of success in this pursuit, he spoke these words in tribute to Max Planck: "I believe . . . that one of the strongest motives that leads men to art and science is the escape from everyday life with its painful crudity and hopeless dreariness, from the fetters of one's own everyday desires . . . a finely tempered nature longs to escape from personal life into the world of objective perception and thought."

In Beethoven's Ninth Symphony, after three movements of breathtaking beauty, the composer interrupts the final movement with the baritone's thunderous introduction to the chorale section, "Oh friends, not these notes" (*nicht diese töne*). As spectacular as his previous creations had been, the composer was searching for something different, something better. Similarly, Einstein could not hear the musicality of his quantum creations, and would spend the rest of his life in search of the final movement that would bring his atomic symphony to a harmonious resolution.

APPENDIX 1: THE PHYSICISTS

In order of appearance:

Max Planck (1858–1947): German theorist, expert on thermodynamics, and Nobel laureate (1918) who introduced the first quantum ideas and his famous constant, h, in order to explain the blackbody radiation law.

Wilhelm Wien (1864–1928): German Nobel laureate (1911) who did the first important theoretical work on the blackbody radiation law, leading in 1896 to Wien's law, which is now known to be an approximation to the correct Planck law.

Heinrich Weber (1843–1912): German experimentalist and researcher in thermodynamics. He was head of the Physics Department at the Zurich Polytechnic when Einstein was a student and clashed with him. His measurements of the temperature variation of specific heats influenced Einstein's 1907 quantum theory of specific heat.

Marcel Grossmann (1878–1936): Swiss mathematician and classmate of Einstein's at the Zurich Poly. His family connections played the key role in Einstein's receiving the patent office job in Bern. Later, in 1913, he became a professor at ETH and collaborated with Einstein on a fundamental paper in General Relativity Theory.

Mileva Maric (1875–1948): Promising physics student who became Einstein's first wife and, after failing to obtain her diploma, did not pursue a career in physics.

Sir Isaac Newton (1642–1726): Founder of classical mechanics through Newton's three laws and the invention of calculus. If you are reading this book you know who he is.

Michael Faraday (1791–1867): English scientist whose experiments led to the concept of electric and magnetic fields; he was greatly admired by Einstein.

James Clerk Maxwell (1831–1879): Scottish theoretical physicist who first found the complete equations of classical electromagnetism, which are named for him. He was also a pioneer of statistical mechanics.

Ludwig Boltzmann (1844–1906): Austrian theoretical physicist who, along with Maxwell and Gibbs, founded the discipline of statistical mechanics. He discovered the fundamental microscopic law of entropy, $S = k \log W$, where k is a fundamental constant of nature known as Boltzmann's constant.

Josiah Willard Gibbs (1839–1903): American physicist and mathematician who, along with Boltzmann and Maxwell, founded statistical mechanics.

Hendrick Antoon Lorentz (1853–1928): Dutch theorist and Nobel laureate (1902) who initially doubted the validity of the Planck law. He became a close friend of, and father figure to Einstein, who regarded him as the greatest thinker he had ever met.

Lord Rayleigh (1842–1919): English mathematical physicist and Nobel laureate (1904); he was an expert on wave theory, particularly acoustics. He proposed the Rayleigh-Jeans law based on classical statistical mechanics, which leads to the incorrect prediction of the ultraviolet catastrophe.

James Jeans (1877–1946): English theoretical physicist and astronomer who contributed to, and also championed, the Rayleigh-Jeans law.

Svante Arrhenius (1859–1927): Swedish physicist and physical chemist, Nobel laureate (1903) for his work on electrolysis, who influenced the establishment and awarding of the Nobel Prizes in Physics and in Chemistry.

Arnold Sommerfeld (1868–1951): German theoretical physicist and leader in the development of the Bohr-Sommerfeld approach to the quantum theory of atoms.

Johannes Stark (1874–1957): German experimental physicist and Nobel laureate (1919), expert on the photoelectric effect, who led the anti-Semitic physics movement under the Nazis.

Walther Nernst (1864–1941): German physical chemist, inventor, and Nobel laureate (1920); he proposed the Third Law of thermodynamics and recruited Einstein to Berlin.

Niels Bohr (1885–1962): Danish physicist and Nobel laureate (1922), who proposed the first successful quantum theory of the atom and played a leading role in interpreting the final form of quantum mechanics.

Ernest Rutherford (1871–1937): New Zealand–born British physicist, Nobel laureate in chemistry (1908). His experiments revealed the nuclear structure of the atom.

Paul Ehrenfest (1880–1933): Austrian Jewish physicist who became professor in Leiden and a significant contributor to quantum theory. He was one of Einstein's closest friends.

Arthur Eddington (1882–1944): English astrophysicist who led the eclipse expedition that confirmed Einstein's theory of general relativity.

Satyendra Nath Bose (1894–1974): Indian theoretical physicist who first proposed the correct statistical method of treating quantum particles as indistinguishable in his paper on photons, sent to Einstein in 1924.

Erwin Schrödinger (1887–1961): Austrian theoretical physicist and Nobel laureate (1933). He invented the wave equation approach to quantum mechanics, which is the main approach used in modern physics.

Duke Louis de Broglie (1892–1987): French theoretical physicist and Nobel laureate (1929). He proposed the idea of matter waves, the complement to Einstein's notion of quanta of light, and influenced Einstein's work on the quantum gas of atoms.

Max Born (1882–1970): German Jewish theoretical physicist and Nobel laureate (1954). He played a major role in formalizing the Heisenberg approach to quantum mechanics, known as matrix mechanics, as well as in providing the probabilistic interpretation of Schrödinger waves. He was one of Einstein's closest friends.

Werner Heisenberg (1901–1976): German theoretical physicist and Nobel laureate (1932). He invented the first correct formulation of modern quantum mechanics, matrix mechanics, in 1925. Two years later he proposed his uncertainty principle.

Wolfgang Pauli (1900–1958): Austrian theoretical physicist and Nobel laureate (1945). A brilliant but caustic personality, he discovered that electrons do not obey Bose statistics, because only one electron can occupy a given quantum state, the Pauli exclusion principle.

APPENDIX 2: THE THREE THERMAL RADIATION LAWS

OVERVIEW

All objects emit electromagnetic radiation because they contain some amount of thermal energy that excites some of their atoms and molecules, at any one time, to higher energy states. These excited atoms and molecules then emit radiation (photons) and fall back down to their lowest (ground) state, while others are continually being excited, which maintains energy balance (thermal equilibrium). The amount of thermal energy an object has increases with its temperature, and so it emits more energetic, higher-frequency radiation as its temperature increases. But thermal radiation, unlike radio waves or laser light, for example, is not emitted at a single frequency; it is emitted over a broad band of frequencies, with the most radiation coming out at a specific frequency (the "peak" of the radiation curve), which depends on the temperature.

The key question challenging physicists circa 1900 was how much energy in the form of radiation is emitted in each band of frequencies for a perfect emitter at a given temperature. Because of a principle of thermodynamics known as Kirchoff's law, a perfect emitter must also be a perfect absorber of radiation (i.e., a perfectly black object), called a blackbody, and the radiation it emits is called blackbody radiation. That doesn't necessarily mean it will appear black to the eye; if it is heated to a sufficiently high temperature, it will glow at optical frequencies, which are visible to the eye. Thus the holy grail for the physics of heat at the time was the determination of the universal mathematical formula describing the energy of heat radiation (per unit volume) in a given frequency band for a given temperature; we will call this formula "the thermal radiation law." There were three radiations laws of historical importance.

1. The Planck law: This law was proposed and then derived by Max Planck in the fall of 1900, when it became clear that the Wien law failed. In order to justify this new law, Planck had to introduce the concept of quantization of energy (although he did not put it that way), which set the quantum revolution in motion. It required the introduction of Planck's constant, h, which appears in the radiation law and in the fundamental relation $\varepsilon = h v$, relating the allowed increments of the energy, E, of a vibrating molecule to its frequency of vibration, v. His law fit the experimental data very well and still does to this day. It is the correct radiation law according to modern physics.

2. The Wien law: This law was proposed by Wilhelm Wien in the early 1890s and was believed by many, including Planck, to be correct until 1900. It turns out to be an approximation to the correct Planck law that works well when one looks at frequencies that are higher than the peak frequency of the Planck law for a given temperature. For the achievable temperatures for blackbody experiments at the time of these first measurements, the higher frequencies were at or near the visible portion of the electromagnetic spectrum, and hence more easily measured, than those below the peak frequency, which were at infrared wavelengths.

3. The Rayleigh-Jeans law: A version of this law was proposed provisionally by Lord Rayleigh in 1900, and it turns out to be an approximation to the correct Planck law at low frequencies, that is, well below the peak frequency predicted by the Planck law. It was then proposed in its correct form by Rayleigh, with input from James Jean, in 1905; at roughly the same time it was derived but then rejected by Albert Einstein. Einstein rejected because it led to the "ultraviolet catastrophe." This ominous descriptor was invented by the physicist Paul Ehrenfest because the Rayleigh-Jeans law predicts that the total energy in thermal radiation should be infinite. Einstein believed this property ruled out the law, whereas Jeans argued for a loophole that kept the theory in play until roughly 1911, when the Planck law became universally accepted.

MATHEMATICAL STATEMENT OF THE RADIATION LAWS

This section assumes that one knows the properties of the exponential function, e^x. Here the letter e stands for the irrational number that is the base of the natural logarithm. The object to be calculated to describe thermal radiation is the so-called spectral energy density of radiation, $\rho(v,T)$, which describes the energy of radiation emitted by a blackbody in a small interval centered at frequency v for an object at absolute temperature T. According to Planck the correct form of this function is

$$(1)\ \rho(v,T) = \frac{(8\pi v^2/c^3)hv}{e^{hv/kT}-1}\ \text{(Planck law)},$$

where h is Planck's constant, k is Boltzmann's constant, and c is the speed of light. Note that the factor in parentheses in the numerator is the one that Bose mentioned in his letter to Einstein in 1924, which contained his new derivation of the radiation law.

The exponential function gets very large compared to 1 when its argument, in this case the ratio hv/kT, is larger than 1 (i.e., when $hv > kT$). When this is the case, one can neglect the term –1 in the denominator of the Planck law, and one finds the approximate form of the radiation law

$$(2)\ \rho(v,T) \approx (8\pi hv^3/c^3)e^{-hv/kT}\ \text{(Wien law)}.$$

This is the form of the radiation law proposed by Wilhem Wien, which we now know is a good approximation for frequencies $v \gg kT/h$. This defines what is meant by "high frequency"; it is a relative term, and it depends on the temperature of the blackbody. Visible light is considered high frequency at the temperature of the earth's surface but not at the temperature of the sun's surface.

In the other limit, of low frequencies, when the $v \ll kT/h$, the exponential function becomes close to the value 1. In fact $e^x \approx 1 + x$ (here x stands for hv/kT), and if we put this into equation (1) for the Planck law, a factor of hv cancels in between the numerator and denominator, and with it all trace of the quantization of energy inherent in the Planck law vanishes. The resulting formula is

(3) $\rho(v,T) \approx (8\pi v^2/c^3)kT$ (Rayleigh-Jeans law).

Note that this approximation to the thermal radiation law shows that for a given frequency, the amount of energy is proportional to the temperature. This is the experimental clue that led Planck to the correct radiation law. However, for a given temperature the energy density is also proportional to the square of the frequency. If this were the correct radiation law for *all* frequencies, one would reach the absurd conclusion that the energy density for high frequencies tends to infinity. This is the ultraviolet catastrophe that Einstein objected to in 1905.

The graph in figure A2.1 compares the three radiation laws.

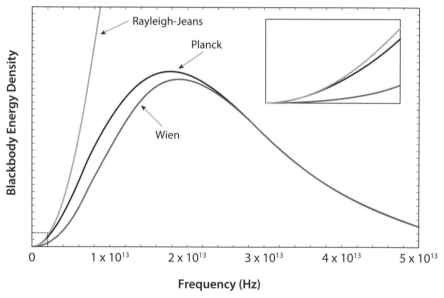

FIGURE A2.1. Graph of the three proposed Radiation Laws, with energy density on the vertical axis and frequency on the horizontal axis, for a blackbody at room temperature. The lightest gray curve, which simply increases to infinity, is the Rayleigh-Jeans Law; note that it agrees with the Planck Law (black curve) for very low frequencies. The black curve is the Planck Law with a peak and then a decay, but with a more rapid rise at low frequencies than the Wien Law (gray). The Wien Law disagrees substantially with the Planck Law below the peak, but agrees very well with it above the peak. Inset is a blow up of the little dashed box in the graph at low frequencies, to show the excellent agreement between the Planck and Rayleigh-Jeans Law, and large disagreement with Wien's Law that was first found experimentally circa 1900. Courtsey of Alex Cerjan.

NOTES

BOOK EPIGRAPH

Page

vi   *Science as something already in existence*: John Stachel, "Einstein and the Quantum," in *Einstein from B to Z*, vol. 9 (Boston: Birkhauser, 2002), 373.

INTRODUCTION. A HUNDRED TIMES MORE THAN RELATIVITY THEORY

Page

3   *I have thought a hundred times*: Abraham Pais, *Subtle Is the Lord* (Oxford: Oxford University Press, 2005), 9.

CHAPTER 1. "AN ACT OF DESPERATION"

Page

7   *It would be edifying if*: Albert Einstein, "Max Planck as a Scientist," in *The Collected Papers of Albert Einstein*, trans. Anna Beck (Princeton: Princeton University Press, 1987–2009), vol. 4, doc. 23, p. 273. All citations are to the English editions of the papers and will be abbreviated *CPAE*.

8   *my decision to devote myself*: Max Planck, *Scientific Biography and Other Papers*, trans. Frank Gaynor (New York: Philosophical Library, 1949), 13.

10   *the limits of validity*: Thomas Kuhn, *Black-Body Theory and the Quantum Discontinuity, 1894–1912* (Chicago: University of Chicago Press, 1978), 89.

10   *the only physical theory* : P. A. Schilpp, ed., *Albert Einstein: Philosopher-Scientist* (La Salle: Open Court, 1970), 33.

10   *he is the model of the theorist*: Planck to Wien, 27 February 1909, in J. L. Heilbron, *Dilemmas of an Upright Man: Max Planck as Spokesman for German Science* (Berkeley: University of California Press, 1986), 8.

11   *The interesting results of long wavelength*: Max Planck, "On an Improvement of Wien's Equation for the Spectrum" [original article no. 1 in References], reprinted in Haar, *The Old Quantum Theory*, 79.

11   *by nature . . . peacefully inclined*: Planck to Robert William Woods, 7 October 1931, in Armin Hermann, *The Genesis of Quantum Theory (1899–1913)* (Cambridge, MA: MIT Press, 1971), 23.

12   *I should therefore be permitted*: Planck, "On an Improvement," 81.

12   *even if the absolutely precise*: Planck, *Scientific Autobiography and Other Papers* (New York: Philosophical Library, 1949), 41.

12   *some weeks of the most strenuous*: Max Planck, Nobel Lecture, p. 4, http://nobel prize.org/physics/laureates/1918/planck-lecture.html, accessed May 14, 2006.

12   *We consider, however*: Max Planck, "On the Theory of the Energy Distribution Law of the Normal Spectrum" [original article no. 2 in References], reprinted in translation in Haar, *The Old Quantum Theory*, 84.

13   *spooky action at a distance*: Einstein to Schrödinger, 19 June 1935, in Walter Isaacson, *Einstein: His Life and Universe* (New York: Simon & Schuster, 2001), 454.

13   *an act of desperation*: Hermann, *Genesis of Quantum Theory*, 23.

13   *I considered the [quantum hypothesis]*: Planck to Woods, ibid., 24.

CHAPTER 2. THE IMPUDENT SWABIAN

Page

15   *neurasthenic exhaustion*: Albrecht Folsing, *Albert Einstein: A Biography*, trans. and abrid. Ewald Osers (New York: Penguin Press, 1998), 30.

15   *assured them most resolutely*: Ibid., 32.

15   *prodigy*: Ibid., 36.

16   *Weber lectured on heat*: Einstein to Mileva Maric, 16 February 1898, *CPAE*, vol. 1, doc. 39, p. 123.

17   *[Weber's] lectures were outstanding*: Carl Seelig, *Albert Einstein: A Documentary Biography*, trans. Mervyn Savill (London: Staples Press, 1956), 29.

17   *Einstein's hopes of learning*: Ibid.

17   *This is the first lecture on mathematical*: Folsing, *Albert Einstein*, 58.

17   *[Einstein] would never get anywhere*: Ibid., 27.

17   *It is pretty much the same*: Seelig, *Albert Einstein*, 19.

18   *A cold wind of skepticism*: Ibid., 14.

18   *holy zeal*: Folsing, *Albert Einstein*, 61.

18   *Herr Weber . . . Herr Professor*: Seelig, *Albert Einstein*, 30.

18   *there is no lack of eagerness*: Ibid., 40.

19   *You're a very clever boy*: Ibid., 30.

19   *regulation paper*: Ibid.

20   *tremendous surprise*: Ibid., 28.

CHAPTER 3. THE GYPSY LIFE

Page

21   *Was a model student*: Seelig, *Albert Einstein*, 34.

21   *vagabond existence*: Folsing, *Albert Einstein*, 52–53.

21   *his impulsive and upright nature*: Seelig, *Albert Einstein*, 38.

22   *you must go all the way*: Ibid., 39.

22   *What the one teaches is not right*: Ibid., 40.

22   *I never saw a trace*: Ibid., 43.

23   *Einstein is 5 ft 9*: Seelig, *Albert Einstein*, 58.

23   *He had the kind of male beauty*: Folsing, *Albert Einstein*, 44.

24   *It's now been quite a while*: Einstein to Maja Einstein, 1898, *CPAE*, vol. 1, doc. 38, p. 123.

24   *I am a cheerful fellow*: Einstein to Alfred Stern, 3 May, 1901, *CPAE*, vol. 1, doc. 104, p. 168.

24   *My broodings about radiation*: Einstein to Mileva Maric, after 28 November 1989, *CPAE*, vol. 1, doc. 43, p. 126.

25   *If only you would be again*: Einstein to Mileva Maric, 10 August 1899, *CPAE*, vol. 1, doc. 52, p. 131.

25   *Mama and Papa are very phlegmatic*: Einstein to Mileva Maric, 9 August 1900, *CPAE*, vol. 1, doc. 71, p. 144.

25   *You too don't like the philistine*: Einstein to Mileva Maric, 3 October 1900, *CPAE*, vol. 1, doc. 79, p. 152.

25   *One must shun*: Einstein to Mileva Maric, 9 August 1900, *CPAE*, vol. 1, doc. 71, p. 144.

25   *Neither of us have gotten a job*: Einstein to Helene Kaufler, 11 October 1900, *CPAE*, vol. 1, doc. 81, p. 153.

CHAPTER 4. TWO PILLARS OF WISDOM

Page

26   *The man loved mysterious Nature*: Harry Woolf, ed., quoted by Freeman Dyson, *Some Strangeness in Proportion: Einstein Centennial* (Reading: Addison-Wesley, 1980), 376.

26   *About Max Planck's studies*: Einstein to Mileva Maric, 4 April 1901, *CPAE*, vol. 1, doc. 96, p. 162.

26   *soon I will have honored*: Ibid., 163.

27   *Maybe his newest theory*: Einstein to Mileva Maric, 10 April 1901, *CPAE*, vol. 1, doc. 97, p. 163.

27   *To find facts which would attest*: Albert Einstein, "Autobiographical Notes," in Schilpp, ed., *Albert Einstein*, 46.

29   *In the beginning—if such a thing*: Seelig, *Albert Einstein*, 32.

29   *We may regard the present*: David Lindley, *Uncertainty: Einstein, Bohr, and the Struggle for the Soul of Science* (New York: Doubleday, 2007), 22.

31   *what's the go o' that*: Ivan Tolstoy, *James Clerk Maxwell* (Edinburgh: Canongate, 1981), 14.

31   *He was the one acknowledged genius*: Ibid., 55.

31   *Happy is the man*: Ibid., 63.

32   *I was sent for to London*: Ibid., 153.

33   *was it God that wrote*: Ibid., 126.

34   *From the long view*: Richard Feynman, *The Feynman Lectures on Physics*, vol. 2 (Reading: Addison-Wesley, 1964), 1–6.

34   *I have also a paper afloat*: Tolstoy, *Maxwell*, 126.

34   *I have been thinking*: Ibid., 169.

34   *[the purely mechanical world . . .]*: Pais, *Subtle Is the Lord*, 319.

35   *Imagine his feelings when*: Ibid.

35   *to few men in the world*: Tolstoy, *Maxwell*, 2.

35   *No one has ever seen*: J. C. Maxwell, *The Scientific Papers of James Clerk Maxwell*, ed. W. D. Niven, vol. 2 (Dover, NY: Dover Publications, 1965), 376–377; Address to Royal Society, April 1873, *Nature*, September 1873; James Clerk Maxwell, "Molecules," Victorian Web, http://www.victorianweb.org/science/maxwell/molecules.html, accessed July 20, 2008.

CHAPTER 5. THE PERFECT INSTRUMENTS OF THE CREATOR

Page

36   *The Boltzmann is magnificent*: Einstein to Maric, September 1900, *CPAE*, vol. 1, doc. 75, p. 149.

36   *By the study of Boltzmann*: Maxwell to P. G. Tait, August 1873, in Pais, *Subtle Is the Lord*, 66.

36   *Boltzmann . . . is not easy*: Seelig, *Albert Einstein*, 104.

37   *An atom is a body*: Maxwell, *Scientific Papers*, 376–377; Address to Royal Society, April 1873; Maxwell, "Molecules," 1.

37   *Our business this evening*: Ibid., 2.

38   *The most important consequence*: Ibid., 3.

39   *If all these molecules were flying*: Ibid., 4.

39   *Lucretius . . . tells us*: Ibid., 7.

40   *The modern atomists have*: Ibid.

40   *The equations of dynamics*: Ibid., 8.

42   *Natural causes, as we know*: Ibid., 9.

CHAPTER 6. MORE HEAT THAN LIGHT

Page

44   *I have again made the acquaintance*: Einstein to Jost Winteler, 8 July 1901, in Folsing, *Albert Einstein*, 86.

44   *the Valiant Swabian is not afraid*: Einstein to Mileva Maric, 15 April 1901, *CPAE*, vol. 1, doc. 101, p. 166.

45   *As for science, I have a few*: Einstein to Marcel Grossman, 14 April 1901, *CPAE*, vol. 1, doc. 100, p. 165.

46   *my worthless first two papers*: Einstein to Johannes Stark, 7 December 1907, *CPAE*, vol. 2, doc. 66, p. 46.

46   *A letter from Marcelius*: Einstein to Mileva Maric, 12 December 1901, *CPAE*, vol. 1, doc. 127, p. 186.

47   *out of consideration for*: Isaacson, *Einstein*, 72.

47   *The situation with the private lessons*: Einstein to Mileva Maric, November 1901, *CPAE*, vol. 1, doc. 136, p. 181.

47   *It is true . . . that it is very nice*: Einstein to Mileva Maric, 17 February 1902, *CPAE*, vol. 1, doc. 137, p. 193.

47   *testifying to great poverty*: Folsing, *Albert Einstein*, 97.

47   *annoying business of starving*: Einstein to Mileva Maric, May 1901, *CPAE*, vol. 1, doc. 110, p. 173.

48   *leads a very pleasant*: Einstein to Michele Besso, 22 January 1903, *CPAE*, vol. 2, doc. 5, p. 7.

49   *the resemblance is downright*: Folsing, *Albert Einstein*, 110.

49   *not have published those papers*: Ibid., 109.

50   *[Hendrik] Lorentz, I never met*: Pais, *Subtle Is the Lord*, 73.

CHAPTER 7. DIFFICULT COUNTING

Page

55   *The Second Law of Thermodynamics has*: Maxwell to Rayleigh, in James Gerald Crowther, *Scientific Types* (New York: Dufour, 1970), 67.

55   *lively beings incapable of doing work*: Tolstoy, *Maxwell*, 136.

55   *to show that the Second Law*: Ibid., 137.

56   *I derive an expression for the entropy*: Albert Einstein, "On the General Molecular Theory of Heat" [original article no. 5 in References], reprinted in *CPAE*, vol. 2, doc. 5, p. 68.

59   *an act of desperation*: Hermann, *Genesis of Quantum Theory*, 23.

59   *We consider, however*: Planck, "On the Theory of the Energy Distribution Law," 84.

61   *Dividing E by ε*: Ibid.

CHAPTER 8. THOSE FABULOUS MOLECULES

Page

63   *those happy Bernese years*: Ann. M. Hentschel and Gerd Grasshoff, *Albert Einstein: Those Happy Bernese Years* (Bern: Staempfli, 2005), title page.

63   *struck by the extraordinary brilliance*: Isaacson, *Einstein*, 80.

63   *laughed so much*: Roger Highfield and Paul Carter, *The Private Lives of Albert Einstein* (London: Faber & Faber, 1993), 102.

65 *To the immortal Olympia Academy*: Einstein to Solovine and Habicht, 3 April 1953, in Banesh Hoffmann with the collaboration of Helen Dukas, *Albert Einstein: Creator and Rebel* (New York: Viking Press, 1972), 244.

66 *completely useless . . . an awful schlemiel . . . [Besso] has*: Isaacson, *Albert Einstein*, 62.

66 *could not have found a better*: Seelig, *Albert Einstein*, 71.

66 *Dear Habicht, such a solemn air*: Einstein to Habicht, 18 or 25 May 1905, *CPAE*, vol. 5, doc. 27, p. 20.

CHAPTER 9. TRIPPING THE LIGHT HEURISTIC

Page

70 *The wave [Maxwell] theory*: Albert Einstein, "On a Heuristic Point of View concerning the Production and Transformation of Light" [original article no. 6 in References], reprinted in *CPAE*, vol. 2, doc. 14, p. 86.

70 *the most revolutionary sentence*: Folsing, *Albert Einstein*, 143.

72 *within a few hours* (quoted in footnote): Ibid., 71.

73 *On my side, I have been*: Besso to Einstein, 17 January 1928, *Einstein Besso Correspondance, 1903–1955*, trans. Pierre Speziali (Paris: Hermann, 1972), 225.

73 *Planck's formula for ρ(v)*: Einstein, "On a Heuristic Point of View," 90.

76 *Morbid dread of mistakes* (quoted in footnote): Crowther, *Scientific Types*, 58.

77 *it makes no difference*: Ibid., 98.

79 *As far as I can see*: Einstein, "On a Heuristic Point of View," 101.

CHAPTER 10. ENTERTAINING THE CONTRADICTION

Page

80 *I do not seek the meaning*: Planck to Einstein, 6 July 1907, *CPAE*, vol. 5, doc. 47, p. 31.

80 *It is a real pity*: Lorentz to Einstein, 6 May 1909, *CPAE*, vol. 5, doc. 153, p. 107.

83 *Take electricity out*: Folsing, *Albert Einstein*, 159.

84 *But such a thing does not seem to exist*: Ibid., 165.

84 *The introduction of a 'light ether'*: Einstein, "On the Electrodynamics of Moving Bodies" [original article no. 7 in References], reprinted in *CPAE*, vol. 2, doc. 23, p. 141.

85 *one can obtain a satisfactory theory*: Einstein, "On the Development of Our Views concerning the Nature and Constitution of Radiation" [original article no. 11 in References], reprinted in *CPAE*, vol. 2, doc. 60, p. 383.

CHAPTER 11. STALKING THE PLANCK

Page

86 *The three of us are fine*: Einstein to Solovine, 27 April 1906, *CPAE*, vol. 5, doc. 36, p. 25.

86   *a respectable federal ink pisser*: Einstein to Alfred Schnauder, 5 January–11 May 1907, *CPAE*, vol. 5, doc. 43, p. 28.

88   *It is largely due to*: Folsing, *Albert Einstein*, 201.

89   *simply a systematic . . . There has been a false*: Ibid., 211.

89   *but more urgent than*: Planck to Einstein, 6 July 1907, *CPAE*, vol. 5, doc. 47, p. 31.

90   *sufficient to account for all*: Einstein, "On a Heuristic Point of View," 90.

90   *In a study published last year*: Albert Einstein, "On the Theory of Light Production and Light Absorption" [original article no. 8 in References], reprinted in *CPAE*, vol. 2, doc. 34, p. 192.

92   *Hence we must view the following proposition*: Ibid., 195.

92   *For if the energy of a resonator . . . In my opinion*: Ibid., 196.

CHAPTER 12. CALAMITY JEANS

Page

95   *the domain of natural sciences*: J. J. O'Connor and E. F. Roberston, "Presidential Address to the British Association in Montreal in 1884," last modified February 2005, accessed 20 February 2009, http://www-history.mcs.st-andrews .ac.uk./Biographies/Rayleigh.html.

95   *little more than a conjecture*: Lord Rayleigh, "Remarks upon the Law of Complete Radiation" [original article no. 3 in References], reprinted in *Scientific Papers by Lord Rayleigh*, vol. 6, doc. 260, p. 483.

97   *what would appear to be wanted*: Lord Rayleigh, "The Law of Partition of Kinetic Energy" [original article no. 5 in References], reprinted in *Scientific Papers by Lord Rayleigh*, vol. 6, doc. 253, p. 451.

98   *the question is one to be settled*: Rayleigh, "Remarks upon the Law," 484.

98   *although for some reason*: Ibid.

99   *If an interaction between aether and matter*: Hermann, *Genesis of Quantum Theory*, 33.

100 *We can now trace the course*: Ibid.

102 *he is the model of the theorist*: Planck to Wien, 27 February 1909, in Heilbron, *Dilemmas of an Upright Man*, 8.

CHAPTER 13. FROZEN VIBRATIONS

Page

103 *The whole thing started*: "Peter J. W. Debye: An Interview," *Science*, vol. 145, p. 554 (1964).

103 *one is allowed to infer*: Pais, *Subtle Is the Lord*, 390.

104 *difficulty and expense*: Ibid.

106 *through the diamond experiment*: Diana Kormos Barkan, *Walther Nernst and the Transition to Modern Physical Science* (Cambridge: Cambridge University Press, 1999), 163.

106 *sunk back into the sea*: Einstein to Mileva Maric, April 4 1901, *CPAE*, vol. 1, doc. 96, p. 162.

107 *For although one has thought before*: Albert Einstein, "Planck's Theory of Radiation and the Theory of Specific Heat," *Annalen der Physik*, vol. 22, pp. 180–190 (1907), reprinted in *CPAE*, vol. 2, doc. 38, p. 218.

107 *I believe we must not content ourselves*: Ibid., 218–219.

110 *most certainly there could exist*: A. Einstein, "Correction to my Paper, 'Planck's Theory of Heat Radiation, etc.,'" *CPAE*, vol. 2, doc. 42, p. 233.

CHAPTER 14. PLANCK'S NOBEL NIGHTMARE

Page

111 *The two constants*: Hermann, *Genesis of Quantum Theory*, 11.

112 *sacrifice him*: Patrick Coffey, *Cathedrals of Science* (Oxford: Oxford University Press, 2008), 3.

113 *[Arrhenius] is somewhat corpulent*: Ibid., 17.

113 *pronounced hostility toward atomism*: Elisabeth Crawford, "Arrhenius, the Atomic Hypothesis, and the 1908 Nobel Prizes in Physics and Chemistry," *Isis*, vol. 75, p. 510 (1984).

114 *he had seen many transmutations* (quoted in footnote): Pierre Marage and Gregoire Wallenborn, eds., *The Solvay Councils and the Birth of Modern Physics*, Science Networks, vol. 22 (Basel: Birkauser Verlag, 1999), 118.

115 *To conclude, I may point to*: Planck, "On the Theory of the Energy Distribution Law," 82–90.

115 *If the theory is at all correct*: Ibid.

116 *concern for a physics*: Heilbron, *Dilemmas of an Upright Man*, 54.

116 *I could derive some satisfaction*: Hermann, *Genesis of Quantum Theory*, 21.

117 *made it extremely plausible*: Ibid., 362.

117 *it is very far from being*: Bengt Nagel, "The Discussion concerning the Nobel Prize for Max Planck," in *Science, Technology and Society in the Time of Alfred Nobel: Nobel Symposium 52*, ed. C. G. Bernhard, E. Crawford, and P. Sorbom (New York: Pergamon Press, 1982), 361.

118 *[if true] I presume*: Crawford, "Arrhenius, the Atomic Hypothesis, and the 1908 Nobel Prizes," 519.

118 *Einstein simply postulates*: Folsing, *Albert Einstein*, 215.

118 *Everything which emanated*: H. A. Lorentz, *Impressions of His Life and Work*, ed. G. L. de Haas-Lorentz (Amsterdam: North-Holland, 1957), 8.

119 *Whatever was accepted by Lorentz*: Hermann, *Genesis of Quantum Theory*, 45.

119 *wrestled continuously*: Lorentz to Wien, 6 June 1908, ibid., 34.

119 *The theory of Planck is the only one*: Ibid., 37.

120 *If we examine the Jeans-Lorentz formula*: Ibid., 40.

120 *I was extremely disappointed*: Wien to Sommerfeld, 18 May 1908, ibid., 39–40.

120 *thus we should really dismiss*: Lorentz to Wien, 6 June 1908, ibid., 41.

121 *a completely new hypothesis*: Nagel, "The Discussion concerning the Nobel Prize for Max Planck, 363.

121 *It is I, along with Phragmen*: Ibid.

CHAPTER 15. JOINING THE UNION

Page

122 *So, now I too am an official*: Einstein to Jakob Laub, 19 May 1909, *CPAE*, vol. 5, doc. 161, p. 120.

122 *I must tell you quite frankly*: Laub to Einstein, 1 May 1908, *CPAE* vol. 5, doc. 91, p. 63.

123 *a few old fogies*: Einstein to Besso, 6 March 1952, ibid.

123 *pigsty*: Einstein to Besso, 7 March 1903, in Folsing, *Albert Einstein*, 228.

123 *I can't understand a word*: Seelig, *Albert Einstein*, 88.

124 *[Einstein] will most likely*: F. Adler to V. Adler, 19 June 1908, in Folsing, *Albert Einstein*, 247.

124 *I was really lucky*: Einstein to Laub, *CPAE*, 19 May, vol. 5, doc. 161, p. 120. I have used the Folsing translation.

125 *impression of a colonel*: "Sommerfeld, Arnold (Johannes Wilhelm)," *Complete Dictionary of Scientific Biography* (2008), Encyclopedia.com, http://www.encyclopedia.com/doc/1G2-2830904080.html, accessed February 4, 2013.

126 *we are now all longing*: Arnold Sommerfeld to Hendrik Lorentz, 26 December 1907, Folsing, *Albert Einstein*, 203.

126 *Your letter made me uncommonly happy*: Einstein to Sommerfeld, 14 January 1908, *CPAE*, vol. 5, doc. 72, p. 5.

126 *I believe that we are still*: Einstein to Sommerfeld, *CPAE*, vol. 5, doc. 72, p. 5.

127 *There can be no doubt*: Albert Einstein, "On the Present Status of the Radiation Problem" [original article no. 10 in References], reprinted in *CPAE*, vol. 2, doc. 56, pp. 360–361.

127 *Though every physicist*: Ibid., 363.

127 *In my opinion, the last*: Ibid., 369.

127 *the fundamental [Maxwell] equation*: Ibid., 374.

128 *I am ceaselessly concerned*: Einstein to Laub, 17 May 1909, *CPAE*, vol. 2, doc. 160, p. 119.

CHAPTER 16. CREATIVE FUSION

Page

129 *I am very sorry*: Einstein to George Meyer, 7 June 1909, *CPAE*, vol. 2, doc. 166, p. 167.

129 *Your postcard made me*: Einstein to Anna Meyer-Schmid, 12 May 1909, *CPAE*, vol. 5, doc. 154, p. 115.

130   *With that kind of fame*: Isaacson, *Einstein*, 154.

130   *That's not true, Herr Einstein*: Seelig, *Albert Einstein*, 92.

130   *my lectures keep me very busy*: Einstein to Besso, 17 November 1909, *CPAE*, vol. 5, doc. 187, p. 140.

131   *I am sending you a short paper*: Einstein to Lorentz, 30 March 1909, *CPAE*, vol. 5, doc. 146, p. 105.

131   *a real event*: Einstein to Lorentz, 13 April 1909, *CPAE*, vol. 5, doc. 149, p. 106.

132   *their existence [free electrons] in metals*: Lorentz to Einstein, 6 May 1909, *CPAE*, vol. 5, doc. 153, p. 108.

132   *If one regards h*: Ibid., 110.

132   *The individuality of each*: Ibid., 112.

132   *as soon as one makes*: Ibid., 113.

132   *permit me to say*: Ibid., 114.

133   *I am presently carrying on*: Einstein to Laub, 19 May 1909, ibid., 119.

134   *My work on light quanta*: Ibid.

134   *delighted . . . the difficulty of generalizing*: Einstein to Lorentz, 23 May 1909, *CPAE*, vol. 5, doc. 163, p. 122.

134   *I consider it a great blessing*: Ibid.

135   *energy elements . . . play a certain*: Planck to Lorentz, 10 July 1909, in Hermann, *Genesis of Quantum Theory*, 41.

135   *stubbornly opposes*: Einstein to Johannes Stark, 31 July 1909, *CPAE*, vol. 2, doc. 172, p. 129.

135   *The quarrel between Stark and Sommerfeld* (quoted in footnote): Einstein to Laub, 16 March 1910, *CPAE*, vol. 2, doc. 199, p. 148.

136   *one of the turning points*: Folsing, *Einstein*, 257.

136   *Once it has been recognized . . . However, today we must*: Einstein, "On the Development of Our Views," 379.

137   *It is therefore my opinion that the next stage*: Ibid.

137   *Regarding our conception of the structure of light*: Ibid., 386.

138   *to accept Planck's theory*: Ibid., 390.

138   *Isn't it conceivable*: Ibid.

139   *The forces of pressure exerted*: Ibid., 391.

140   *it has not yet been possible*: Ibid., 394.

140   *listening with greatest interest*: "Discussion Following the Lecture: On the Development of Our Views concerning the Nature and Constitution of Radiation" [original article no. 12 in References], reprinted in *CPAE*, vol. 2, doc. 61, pp. 395–396.

140   *It had no [great effect]*: Hermann, *Genesis of Quantum Theory*, 68.

CHAPTER 17. THE IMPORTANCE OF BEING NERNST

Page

141   *I visited Prof. Einstein in Zurich*: W. Nernst to A. Schuster, 10 March 1910, in Barkan, *Walther Nernst*, 183.

141 *Einstein's achievement received its seal*: Folsing, *Albert Einstein*, 257.

142 *the best that any laboratory practitioner*: Coffey, *Cathedrals of Science*, 18.

142 *fishlike mouth*; . . . *in the main, popular* . . . ; and the Perrin anecdote: Robert A. Millikan, "A Great Physicist Passes," *Scientific Monthly*, vol. 54, pp. 84–86 (1942).

142 *his truly amazing scientific instinct*: Albert Einstein, "The Work and Personality of Walther Nernst," *Scientific Monthly*, vol. 54, pp. 195–196 (1942).

144 *How much did you get*: Coffey, *Cathedrals of Science*, 29–30.

144 *versatile, many-faceted*: Barkan, *Walther Nernst*, 148.

144 *Kommerzienrat*: Phillip Frank, *Einstein: His Life and Times*, trans. George Rosen, ed. Schuichi Kusaka (New York: Da Capo Press, 1947), 107.

146 *rather shabby attire*: Seelig, *Albert Einstein*, 100.

146 *I am most interested in the associate professor*: George Bredig to Arrhenius, in Barkan, *Walther Nernst*, 183–4.

146 *made Einstein famous*: Kuhn, *Black-Body Theory*, 215.

147 *I visited Prof. Einstein . . . "a beautiful memory"*: Nernst to A. Schuster, 10 March 1910, in Barkan, *Walther Nernst*, 183.

147 *The specific heat decreases* (quoted in footnote): 17 February 1910, ibid., 167.

148 *I have made inquiries*: W. Nernst postcard, recipient unknown, 31 July 1910, ibid.

148 *For me the theory of quanta*: Einstein to Laub, 16 March 1910, *CPAE*, vol. 5, doc. 199, pp. 148–149.

148 *It seems incontrovertible*: Einstein to Sommerfeld, July 1910, *CPAE*, vol. 5, doc. 211, p. 157.

148 *The crucial point in the whole question*: Ibid.

CHAPTER 18. LAMENTING THE RUINS

Page

149 *As for knowing, nobody knows anything*: Einstein to Zangger, 15 November 1911, *CPAE*, vol. 5, doc. 305, p. 221.

149 *I decided to take as my starting point*: Marage and Wallenborn, eds., *Solvay Councils*, 10–11.

150 *the report he was good enough to send*: Ibid., 95.

151 *chasing after modernity*: Folsing, *Albert Einstein*, 273.

151 *in boldness [surpassing] anything*: Folsing, *Albert Einstein*, 271.

151 *great theoretical papers in the area of thermodynamics* (quoted in footnote): Emil Fischer to Einstein, October–November 1910, *CPAE*, vol. 5, doc. 230, p. 165.

152 *the air is full of soot*: Einstein to Hans Tanner, 24 April 1911, *CPAE*, vol. 5, doc. 265, p. 186.

152 *semi-barbaric*: Einstein to Zangger, 20 September 1911, *CPAE*, vol. 5, doc. 286, p. 207.

152 *a good thing for the Polytechnic*: Einstein to Zangger, date unknown (after 5 June 1912), Folsing, *Albert Einstein*, 79.

153   *as soon as one makes even the slightest*: Lorentz to Einstein, 6 May 1909, *CPAE*, vol. 5, doc. 153, p. 113.

153   *it has not been possible to formulate*: Einstein, "On the Development of Our Views," 394.

153   *I have not yet arrived at solution*: Einstein to Laub, 31 December 1909, *CPAE*, vol. 5, doc. 196, p. 146.

153   *maybe the electron is not*: Einstein to Sommerfeld, 19 January 1910, *CPAE*, vol. 5, doc. 197, pp. 146–147.

154   *[a] crudely materialistic conception*: Einstein to Sommerfeld, July 1910, *CPAE*, vol. 5, doc. 211, p. 157.

154   *I have not made any progress*: Einstein to Laub, 27 August 1910, *CPAE*, vol. 5, doc. 224, p. 162.

154   *At the moment I am very hopeful*: Einstein to Laub, 4 November 1910, *CPAE*, vol. 5, doc. 231, p. 166.

154   *The solution of the radiation problem*: Einstein to Laub, 11 November 1910, *CPAE*, vol. 5, doc. 233, p. 167.

154   *the riddle of radiation*: Einstein to Laub, 28 December 1910, *CPAE*, vol. 5, doc. 241, p. 141.

155   *the so-called quantum theory of today*: Barkan, *Walther Nernst*, 165.

156   *I no longer ask whether these quanta*: Einstein to Besso, 13 May 1911, *CPAE*, vol. 5, doc. 267, p. 187.

157   *Dear Sir, To all appearances*: Ernest Solvay to Einstein, 9 June 1911, *CPAE*, vol. 5, doc. 269, pp. 190–191.

157   *The relativistic treatment of gravitation*: Einstein to Laub, 10 August 1911, *CPAE*, vol. 5, doc. 275, p. 197.

157   *If my answer is not . . . thorough*: Folsing, *Albert Einstein*, 285.

157   *We stand here before an unsolved puzzle*: Albert Einstein, "On the Present State of the Problem of Specific Heats" [original article no. 13 in References], reprinted in *CPAE*, vol. 2, doc. 26, pp. 419–420.

158   *H. A. Lorentz chaired the conference*: Einstein to Zangger, 7 November 1911, *CPAE*, vol. 5, doc. 303, pp. 219–220.

158   *he is a completely honest man*: Ibid., 220.

158   *In Brussels, too*: Einstein to Besso, 26 December 1911, *CPAE*, vol. 5, doc. 331, p. 241.

158   *the h-disease looks ever more hopeless*: Einstein to Lorentz, 23 November 1911, *CPAE*, vol. 5, doc. 313, p. 227.

158   *I took as my starting point*: Marage and Wallenborn, eds., *Solvay Councils*, 11.

158   *I'm working at full speed*: Einstein to Alfred Stern, 17 March 1912, *CPAE*, vol. 5, doc. 374, p. 275.

158   *I assure you that I have nothing*: Einstein to Sommerfeld, 29 October 1912, *CPAE*, vol. 5, doc. 421, p. 324.

159   *My letter to Einstein proved useless*: Sommerfeld to Hilbert, 1 November 1912, in Hermann, *Genesis of Quantum Theory*, 69.

159   *his sense of humor*: Frank, *Einstein*, 76.

159   *There are the madmen*: Ibid., 98.

CHAPTER 19. A COSMIC INTERLUDE

Page

160   *Scientific endeavors are quite extraordinary*: Einstein to Walter Dallenbach, 31 May 1915, *CPAE*, vol. 8, doc. 87, p. 103.

161   *The question arises whether this statement*: Folsing, *Albert Einstein*, 339.

162   *one of the highest achievements*: Ibid., 444.

162   *probably the greatest scientific discovery*: Isaacson, *Einstein*, 223.

162   *the greatest feat of human thinking*: Ibid., 223–224.

162   *The theory is of incomparable beauty*: Einstein to Zangger, 26 November 1915, *CPAE*, vol. 8, doc. 152, p. 151.

162   *my boldest dreams*: Einstein to Besso, 10 December 1915, *CPAE*, vol. 8, doc. 162, p. 160.

163   *enchanted me*: Einstein to Sommerfeld, 8 February 1916, *CPAE*, vol. 8, doc. 189, p. 192.

163   *among my finest experiences*: Einstein to Sommerfeld, 3 August 1916, *CPAE*, vol. 8, doc. 246, p. 241.

163   *A brilliant idea has dawned*: Einstein to Besso, 11 August 1916, *CPAE*, vol. 8, doc. 250, p. 241.

165   *I can't even begin to tell*: Einstein to Elsa, 30 April 1912, *CPAE*, vol. 5, doc. 389, p. 291.

165   *the chances of my getting a call*: Ibid.

165   *it will not be good*: Einstein to Elsa, 21 May 1912, *CPAE*, vol. 5, doc. 399, p. 300.

166   *tear of joy*: Folsing, *Albert Einstein*, 332.

166   *The undersigned are aware* (quoted in footnote): 12 June 1913, *CPAE*, vol. 5, doc. 445, p. 38.

167   *because it would not make me happy*: Frank, *Einstein*, 109.

167   *I couldn't resist the temptation*: Einstein to Lorentz, 14 August 1913, in Folsing, *Einstein*, 331.

167   *I accepted this odd sinecure*: Einstein to Paul Ehrenfest, undated, 1913, ibid.

167   *howled unceasingly*: Highfield and Carter, *Private Lives*, 164.

167   *In personal respects*: Einstein to Zangger, 7 July 1915, *CPAE*, vol. 8, doc. 94, p. 110.

CHAPTER 20. BOHR'S ATOMIC SONATA

Page

168   *Europe in its madness*: Einstein to Ehrenfest, 19 August 1914, *CPAE*, vol. 8, doc. 34, p. 41.

168   *unworthy of what until now*: Folsing, *Albert Einstein*, 346.

169   *the international catastrophe*: Ibid., 347.

170   *Bohr is different*: William Cropper, *Great Physicists: The Life and Times of Leading Physicists from Galileo to Hawking* (Oxford: Oxford University Press, 2004), 245.

170   *full of characters from all parts*: Ibid.

171   *I got a vivid account*: Marage and Wallenborn, eds., *Solvay Councils*, 110.

173   *I really think you should abbreviate*: Jagdish Mehra and Helmut Rechenberg, *The Historical Development of Quantum Theory*, vol. 1, part 1 (New York: Springer-Verlag, 1982), 190. (Cited hereinafter as Mehra and Rechenberg, *HDQT*.)

173   *in an attempt to explain*: N. Bohr, "On the Constitution of Atoms and Molecules" [original article no. 14 in References], reprinted in Haar, *The Old Quantum Theory*, pp. 132–133.

174   *The general importance*: Ibid., 137.

174   *For me the only difficulty*: Einstein to Lorentz, 23 May 1909, *CPAE*, vol. 5, doc. 163, p. 122.

177   *How much the above interpretation*: Bohr, "On the Constitution," 149.

178   *obviously, we get in this way*: Ibid., 150.

179   *we came to speak of Bohr's theory*: Stachel, *Einstein from B to Z*, 369–370.

179   *I told him [the explanation . . .]*: Ibid.

180   *All my attempts, however, to adapt*: Albert Einstein, "The Advent of Quantum Theory," *Science*, vol. 113, pp. 82–84 (1951).

CHAPTER 21. RELYING ON CHANCE

Page

181   *Fundamental as [the relativity theory] of Einstein*: *CPAE*, vol. 5, doc. 445, pp. 337–338.

181   *Why Planck and I engaged him*: Barkan, *Walther Nernst*, 186.

182   *One cannot seriously believe*: Einstein to Wilhelm Wien, 17 May 1912, *CPAE*, vol. 5, doc. 395, p. 298.

182   *the process of absorption*: Einstein to Michele Besso, 11 September 1911, *CPAE*, vol. 5, doc. 283, p. 204.

183   *Einstein always wracked his brain*: Woolf, ed., *Some Strangeness in Proportion*, 254.

184   *My measurements . . . agree everywhere*: Mehra and Rechenberg, *HDQT*, vol. 1, part 1, p. 220.

184   *Now I believe in relativity theory*: Ibid., footnote 366, p. 231.

184   *a revelation*: Einstein to Sommerfeld, 8 February 1916, *CPAE*, vol. 8, doc. 189, p. 192.

184   *your spectral analyses*: Einstein to Sommerfeld, 3 August 1916, *CPAE*, vol. 8, doc. 246, p. 241.

184   *his derivation was of unparalleled boldness*: Albert Einstein, "Emission and Absorption of Radiation in Quantum Theory" [original article no. 15 in References], reprinted in *CPAE*, vol. 6, doc. 34, p. 212.

186   *We shall distinguish here*: Ibid., 214.

187   *The simplicity of the hypotheses*: Ibid., 215–216.

187 *A brilliant idea has dawned*: Einstein to Besso, 11 August 1916, *CPAE*, vol. 8, doc. 250, p. 243.

188 *it can be demonstrated convincingly*: Einstein to Besso, 24 August 1916, *CPAE*, vol. 8, doc. 251, p. 244.

188 *this derivation deserves attention*: Albert Einstein, "On the Quantum Theory of Radiation" [original article no. 16 in References], reprinted in *CPAE*, vol. 6, doc. 38, pp. 220–221.

188 *The question arises: does the molecule*: Ibid., p. 221.

192 *If we were to modify*: Ibid., p. 232.

192 *[Einstein] himself, in his paper of [1917]*: Werner Heisenberg, *Encounters with Einstein: And Other Essays on People, Places, and Particles*. (Princeton: Princeton University Press, 1989), 115.

192 *the molecule suffers a recoil*: Einstein, "On the Quantum Theory of Radiation," 232.

192 *any such elementary process*: Einstein to Besso, 6 September 1916, *CPAE*, vol. 8, doc. 254, p. 246.

CHAPTER 22. CHAOTIC GHOSTS

Page

193 *I have firmly decided*: Einstein to Elsa, 11 August 1913, *CPAE*, vol. 5, doc. 466, p. 348.

193 *turnip winter*: Thomas Levenson, *Einstein in Berlin* (New York: Bantam Books, 2003), 142.

194 *famous complaint*: Folsing, *Albert Einstein*, 406.

194 *I am quite infirm*: Einstein to Paul Ehrenfest, 14 February 1917, *CPAE*, vol. 8, doc. 298, p. 295.

194 *I have not been working much*: Einstein to Lorentz, 3 April 1917, *CPAE*, vol. 8, doc. 322, p. 313.

194 *raise the necessary superstition*: Einstein to Besso, 8 May 1917, *CPAE*, vol. 8, doc. 335, p. 325.

194 *I am committing myself*: Einstein to Besso, 13 May 1917, *CPAE*, vol. 8, doc. 339, p. 330.

194 *I have come to know*: Einstein to Zangger, 10 March 1917, *CPAE*, vol. 8, doc. 309, p. 299.

194 *My health is quite fair*: Einstein to Zangger, 6 December 1917, *CPAE*, vol. 8, doc. 404, p. 412.

195 *Elsa indefatigably cooking*: Einstein to Besso, 28 June 1918, *CPAE*, vol. 8, doc. 572, p. 598.

196 *The quantum paper I sent*: Einstein to Besso, 9 March 1917, *CPAE*, vol. 8, doc. 306, p. 292.

196 *scientific life has dozed*: Einstein to Zangger, 10 March 1917, *CPAE*, vol. 8, doc. 310, p. 300.

197 *I am convinced that besides*: Einstein to Arnold Sommerfeld, 9 October 1921, *CPAE*, vol. 12, doc. 261, p. 168.

197 *Basic idea: . . . Upon*: Lorentz to Einstein, 13 November, 1921, *CPAE*, vol. 12, doc. 298, p. 184.

198 *quite fond of it*: Abraham Pais, *Inward Bound: Of Matter and Forces in the Physical World* (New York: Clarendon Press, 1986), 259.

198 *It has to be assumed*: Ibid.

201 *yesterday I presented a little thing*: Einstein to Besso, 29 April 1917, *CPAE*, vol. 8, doc. 331, p. 322.

201 *it still remains unsatisfying*: Albert Einstein, "On the Quantum Theorem of Sommerfeld and Epstein" [original article no. 17 in References], reprinted in *CPAE*, vol. 6, doc. 45, p. 436.

202 *we now come to a very essential point*: Ibid., 439.

202 *One notices immediately*: Ibid., 440.

203 *The framing of the quantum conditions*: Erwin Schrödinger, "Quantization as an Eigenvalue Problem, Part II," in *Collected Papers on Wave Mechanics*, 2nd edition, 1928, translated 1978 (New York: Chelsea Publishing, 1978), 17.

CHAPTER 23. FIFTEEN MILLION MINUTES OF FAME

Page

204 *But now I'm just about fed up*: Einstein to Elsa, 8 January 1921, *CPAE*, vol. 12, doc. 12, p. 3.

205 *a Greek drama*: Folsing, *Albert Einstein*, 442–443.

205 *one of the highest achievements*: Ibid., 444.

205 *Lights All Askew*: Digital facsimiles of the original *New York Times* article can be viewed online at, for example: Yeshiva University, Einstein and Yeshiva Digital Exhibit, http://yu.edu/libraries/digital_library/einstein/Einstein11-10-19nyt .pdf, accessed March 31, 2013.

205 *At present every coachman*: Folsing, *Albert Einstein*, 455.

205 *from the intellectual work of a quiet scholar*: Ibid.

206 *the fragmentation of one's intentions*: Einstein to Zangger, 16 January 1921, *CPAE*, vol. 12, doc. 5, p. 9.

207 *Everyone must, from time to time*: Einstein to Max and Hedwig Born, 9 September 1920, in Folsing, *Albert Einstein*, 464.

207 *sound common sense*: Ibid., 465.

207 *[he] has so far achieved nothing*: Ibid.

207 *I feel like a man lying*: Ibid., 463; Einstein interview of August 29, 1920.

208 *It cannot be our task*: Ibid.

208 *there is in me nothing*: Ibid., 494.

208 *The [religious Jewish] community*: Einstein to Rabbi Warschauer, 8 March 1921, *CPAE*, vol. 12, doc. 86, p. 69.

209 *I am not needed for my abilities*: Einstein to Haber, 9 March 1921, *CPAE*, vol. 12, doc. 88, p. 71.

210 *very interesting and quite simple*: Einstein to Born, 22 August 1921, *CPAE*, vol. 12, doc. 211, p. 143.

211 *Thanks to the excellent collaboration*: Einstein to Born, 30 December 1921, *CPAE*, vol. 12, doc. 345, p. 211.

211 *I . . . committed a monumental blunder*: Einstein to Born, beginning of May 1922, in *The Born-Einstein Letters, 1916–1955: Friendship, Politics and Physics in Uncertain Times*, trans. Irene Born (New York: MacMillan, 1971), 68.

211 *I suppose it is a good thing*: Einstein to Ehrenfest, 15 March 1922, in Folsing, *Albert Einstein*, 512.

212 *a downright normal human existence*: Einstein to H. Anschuz-Kaempfe, 12 July 1922, ibid., 521.

212 *I am constantly being warned*: Einstein to M. Solovine, 16 July 1922, ibid., 522.

212 *the opportunity of a prolonged absence*: Einstein to Wilhelm Solf, 20 December 1922, ibid., 525.

212 *it will probably be very desirable*: Arrhenius to Einstein, 18 September 1922, ibid., 535.

212 *quite unable to postpone*: Einstein to Arrhenius, 20 September 1922, ibid.

213 *I spent ten years*: Pais, *Subtle Is the Lord*, 357.

213 *The other great problem*: Einstein recording for Preusssischen Staatsbibliothek, 1924, quoted in M. J. Klein, "No Firm Foundation: Einstein and the Early Quantum Theory," in Woolf, ed., *Some Strangeness in Proportion*, 182.

CHAPTER 24. THE INDIAN COMET

Page

215 *Respected Sir: I have ventured*: John Stachel, "Einstein and Bose," in *Einstein from B to Z*, vol. 9, p. 524.

216 *I do not know sufficient German*: Ibid.

216 *In my opinion Bose's derivation*: S. N. Bose, "Planck's Law and the Light Quantum Hypothesis" [original article no. 18 in References], reprinted in O. Theimer and B. Ram, "The Beginning of Quantum Statistics," *American Journal of Physics*, vol. 44, pp. 1056–1057 (1976), p. 1056.

216 *about 300 words* (quoted in footnote): Jurgen Neffe, *Einstein: A Biography*, trans. Shelley Frisch (New York: Farrar Strauss Giroux, 2007), 372.

217 *a particularly brilliant lot*: Stachel, "Einstein and Bose," 520.

218 *notorious for plain speaking*: K. Wali, "The Man Behind Bose Statistics," *Physics Today*, October 2006, p. 48.

218 *you may have done well in the examination*: Ibid.

218 *Disappointed, I came away*: Ibid.

218 *spent many sleepless nights*: W. Blanpied, "Satyendranath Bose: Co-founder of Quantum Statistics," *American Journal of Physics*, September 1972, p. 1215.

218 *As a teacher*: Mehra and Rechenberg, *HDQT*, vol. 1, part 2, p. 565.

220 *Planck's formula . . . forms the starting point*: Bose, "Planck's Law and the Light Quantum," 1056.

220 *could be deduced only*: Ibid.

220 *However I do not find your objection*: Stachel, "Einstein and Bose," 524.

222 *the fourth and last*: Pais, *Subtle Is the Lord*, 425.

223 *It is now very simple*: Bose, "Planck's Law and the Light Quantum," 1057.

223 *I had no idea*: Jagdish Mehra, "Satyendra Nath Bose," *Biographical Memoirs of Fellows of the Royal Society*, November 1975, p. 128.

224 *[the Bose] derivation is elegant*: Einstein to Ehrenfest, 12 July 1924, in Pais, *Subtle Is the Lord*, 424.

225 *contradicts the generally and rightly accepted*: Albert Einstein, "'Bose's Second Paper: A Conflict with Einstein,' Comment on 'Thermal equilibrium in the radiation field in the presence of matter,' by S. N. Bose," *Physikalische Zeitschrift*, vol. 27, p. 384 (1924), reprinted in O. Theimer and B. Ram, *American Journal of Physics*, vol. 45, pp. 242–246 (1976), p. 246.

225 *solved all problems*: Stachel, "Einstein and Bose," 525.

225 *work under you*: Bose to Einstein, 26 October 1924, ibid.

225 *I am glad I shall have*: Einstein to Bose, 2 November 1924, ibid., 525–526.

226 *because I was a teacher*: Mehra, "Satyendra Nath Bose," 136.

226 *even more than forty years*: Blanpied, "Satyendranath Bose," 1217.

226 *She was very nice . . . I wasn't able*: Mehra, "Satyendra Nath Bose," 136.

226 *the meeting was most interesting*: Ibid., 141.

226 *what the statistics of light quanta*: Ibid.

227 *I have made an honest resolution of*: Stachel, "Einstein and Bose," 527.

227 *your senator McCarthy*: Blanpied, "Satyendranath Bose," 1217.

227 *His indomitable will*: Mehra, "Satyendra Nath Bose," 145.

227 *On my return*: Ibid., 142.

CHAPTER 25. QUANTUM DICE

Page

228 *I . . . am convinced that [God]*: Einstein to Born, 4 December, 1926, in Folsing, *Albert Einstein*, 585.

230 *A quantum theory of the . . . ideal gas*: Albert Einstein, "Quantum Theory of the Monatomic Ideal Gas" [original article no. 20 in References], reprinted in translation in, I. Duck and E.C.G. Sudarshan, eds., *Pauli and the Spin-Statistics Theorem* (Singapore: World Scientific, 1997), 82.

231 *still obscure*: Einstein to Ehrenfest, 12 July 1924, in Stachel, "Einstein and Bose," 533.

231 *the theory is pretty*: Martin J. Klein, "Einstein and Wave-Particle Duality," *The Natural Philosopher*, vol. 3, p. 31 (1964).

231 *The thing with the quantum gas*: Einstein to Ehrenfest, 2 December 1924, in Mehra and Rechenberg, *HDQT*, vol. 5, part 2, p. 384.

235 *When the Bose derivation*: Albert Einstein, "Quantum Theory of the Monatomic Ideal Gas, Part Two" [original article no. 21 in References], reprinted in translation in I. Duck and E.C.G. Sudarshan, eds., *Pauli and the Spin-Statistics Theorem* (Singapore: World Scientific, 1997), 88.

236 *I maintain that in this case*: Ibid., 90.

238 *Ehrenfest and others have reported*: Ibid., 91.

240 *your reproach is not unjustified*: Einstein to Erwin Schrödinger, 28 February 1925, in Mehra and Rechenberg, *HDQT*, vol. 5, part 2, p. 387.

240 *only through your letter*: Schrödinger to Einstein, 3 November 1925, ibid.

CHAPTER 26. THE ROYAL MARRIAGE: $E = mc^2 = h\nu$

Page

241 *I said to myself that classical physics*: Interview of Louis de Broglie, Archives for the History of Quantum Physics Collection, Niels Bohr Library and Archives, American Institute of Physics, College Park, MD, www.aip.org/history/ohilist /LINK, p. 30, translated from French by the author. (Cited hereinafter as de Broglie, AHQP interview.)

241 *A younger brother of*: Einstein to Lorentz, 16 December 1924, in Mehra and Rechenberg, *HDQT*, vol. 1, part 2, p. 604.

241 *a bit strange, but after all*: De Broglie, AHQP interview, 7.

242 *Louis de Broglie's work*: Ioan James, *Remarkable Physicists, from Galileo to Yukawa* (Cambridge: Cambridge University Press, 2004), 311.

242 *at every stage of my life*: B. R. Wheaton, *The Tiger and the Shark: Empirical Roots of Wave-Particle Dualism* (Cambridge: Cambridge University Press, 1983), 274.

243 *this little brother had become*: A. Abragam, "Louis De Broglie," *Biographical Memoirs of Fellows of the Royal Society*, vol. 34, pp. 22–41 (1988), p. 26.

243 *continue and prepare the diploma*: De Broglie, AHQP interview, 30.

243 *I could see that to do that it was necessary*: Ibid.

243 *between several intellectual directions*: Ibid., 29.

243 *The hesitations are over*: Abragam, "Louis De Broglie," 27.

243 *I began to think about quanta*: De Broglie, AHQP interview, 22.

244 *with the ardor of my age*: Mehra and Rechenberg, *HDQT*, vol. 1, part 2, p. 582.

244 *his enthusiasm was returning*: Abragam, "Louis De Broglie," 28.

244 *regretted the interruption*: Ibid.

244 *because that put me*: De Broglie, AHQP interview, 28.

244 *There was really no career*: Mehra and Rechenberg, *HDQT*, vol. 1, part 2, p. 579.

245 *I never had any doubt*: De Broglie, AHQP interview, 22.

245 *must be corpuscular*: Wheaton, *The Tiger and the Shark*, 270.

245 *a number of known results*: Louis de Broglie, "Black Radiation and Light Quanta" [original article no. 23 in References], reprinted in Louis de Broglie and Leon Brillouin, *Selected Papers on Wave Mechanics*, 1–7 (London: Blackie and Sons, 1928), 1.

245 *incomparable insight*: Louis de Broglie, "Studies on the Theory of Quanta," PhD thesis [original article no. 25 in References], 3.

246 *atoms of light*: Ibid., 6.

246 *a mixture of monatomic, diatomic, triatomic* (quoted in footnote): De Broglie, "Black Radiation," 5.

247 *Well, all this must be very similar*: Mehra and Rechenberg, *HDQT*, vol. 1, part 2, p. 586.

247 *developed rapidly in the summer*: De Broglie, AHQP interview, 25.

247 *I got the idea that one had to extend*: Mehra and Rechenberg, *HDQT*, vol. 1, part 2, p. 587.

248 *a meta law of Nature*: De Broglie, PhD thesis, 8.

248 *glides on its wave*: Asim O. Barut, Alwyn van der Merwe, and Jean-Pierre Vigier, eds., *Quantum Space and Time—the Quest Continues: Studies and Essays in Honour of Louis De Broglie, Paul Dirac and Eugene Wigner* (Cambridge: Cambridge University Press, 1984), 18.

249 *a very remarkable geometric interpretation*: Einstein, "Quantum Theory of the Monatomic Ideal Gas, Part Two," 99.

249 *It looks far-fetched*: Abragam, "Louis De Broglie," 30.

249 *read my thesis during the summer*: Louis de Broglie, AHQP interview, 7.

250 *remarkable mastery . . . never has so much. . . . all I can tell you*: Abragam, "Louis De Broglie," 30.

251 *I believe that it is more*: Einstein, "Quantum Theory of the Monatomic Ideal Gas, Part Two," 95.

251 *this oscillating field*: Ibid., 96.

251 *The scientific world of the time*: Klein, "Wave-Particle Duality." 38.

251 *the paper of de Broglie*: Abragam, "Louis De Broglie," 31.

252 *By way of a detour*: Klein, "Wave-Particle Duality," 39.

252 *there was no experimental evidence* (quoted in footnote): Woolf, ed., *Some Strangeness in Proportion*, footnote, p. 471.

253 *uninspiring*: Abragam, "Louis De Broglie," 37.

253 *dry and devoid of passion*: Ibid.

253 *not of the highest intellectual caliber*: Ibid.

253 *Yesterday I read*: Ibid., 35.

CHAPTER 27. THE VIENNESE POLYMATH

Page

254 *Physics does not consist*: Schrödinger to Wilhelm Wien, 25 August 1926, in Walter Moore, *Schrödinger: Life and Thought* (Cambridge: University of Cambridge Press, 1989), 226.

254 *When you began this work*: Schrödinger, *Collected Papers on Wave Mechanics*, preface, v.

254 *I have read your article*: Einstein to Schrödinger, 2 April 1926, in Martin Klein, ed., *Letters on Wave Mechanics* (New York: Philosophical Library, 1967), 3.

254 *the idea of your article*: Einstein to Schrödinger, 16 April 1926, ibid., 24.

255 *In my scientific work*: Moore, *Schrödinger*, 135.

255 *devote himself to philosophy* (quoted in footnote): Mehra and Rechenberg, *HDQT*, vol. 5, part 1, p. 225.

256 *often been accused of flirtatiousness*: Erwin Schrödinger, *What is Life? The Physical Aspect of the Living Cell with Mind and Matter & Autobiographical Sketches* (Cambridge: Cambridge University Press, 1958), 167.

256 *I must refrain from drawing*: Ibid., 184.

256 *who did not wish, in consequence* (quoted in footnote): Moore, *Schrödinger*, 272.

256 *I have trouble with Dirac* (quoted in footnote): Helge S. Kragh, *Dirac: A Scientific Biography* (Cambridge: Cambridge University Press, 1990), 82.

257 *I was a good student*: Ibid., 23.

257 *only find words of praise*: Schrödinger, *What is Life?* 173.

257 *I can't recall a single instance*: Moore, *Schrödinger*, 23.

257 *das ist der Schrödinger . . . this man . . . a fiery spirit*: Ibid., 46.

257 *would translate Homer*: Mehra and Rechenberg, *HDQT*, vol. 5, part 1, p. 67.

257 *hours of leisure*: Ibid., 68.

257 *if it were not for the mathematics*: Moore, *Schrödinger*, 97.

258 *I learnt to appreciate*: Schrödinger, *What is Life?* 169.

258 *his fearlessness and calmness*: Moore, *Schrödinger*, 93.

258 *I was impressed by him*: Mehra and Rechenberg, *HDQT*, vol. 5, part 1, p. 177.

258 *I know it would be easier*: Moore, *Schrödinger*, 131.

259 *No perception in physics*: Schrödinger, *What is Life?* 168.

260 *beautiful work*: Schrödinger to Bohr, 7 February 1921, in Mehra and Rechenberg, *HDQT*, vol. 5, part 1, p. 306.

260 *[your paper] interested me very much*: Bohr to Schrödinger, 15 June 1921, ibid., 307.

261 *originality of [Einstein's] statistical method*: Schrödinger to Einstein, 3 November 1925, in Mehra and Rechenberg, *HDQT*, vol. 5, part 2, p. 387.

261 *in order that two molecules*: Mehra and Rechenberg, *HDQT*, vol. 5, part 1, p. 365.

261 *I have read with great interest*: Einstein to Schrödinger, 26 September 1925, in Mehra and Rechenberg, *HDQT*, vol. 5, part 2, p. 397.

262 *the basic idea is yours*: Schrödinger to Einstein, 3 November 1925, ibid., 398.

262 *since you have performed*: Einstein to Schrödinger, 14 November 1925, ibid.

262 *not even jokingly*: Schrödinger to Einstein, 4 December 1925, ibid., 398–399.

262 *because of it, section 8*: Schrödinger to Einstein, 3 November 1925, ibid., 412.

262 *return to wave theory*: Schrödinger to Alfred Lande, 16 November 1925, ibid., 418.

263 *nothing else but taking seriously*: Klein, "Wave-Particle Duality," 43.

263 *Wave mechanics was born*: Moore, *Schrödinger*, 188.

263 *I see no basic difference* (quoted in footnote): Einstein to Schrödinger, 22 April 1926, in Klein, ed., *Letters on Wave Mechanics*, 25.

263 *did his great work* (quoted in footnote): Pais, *Inward Bound*, 252.

266 *in this paper, I wish*: Schrödinger, *Collected Papers on Wave Mechanics*, 1.

266 *It is, of course, strongly suggested*: Ibid., 9.

266 *Above all, I wish to mention*: Ibid.

266 *My theory was inspired*: Ibid., footnote, p. 46.

267 *What is more magnificent*: Mehra and Rechenberg, *HDQT*, vol. 5, part 1, p. 1.

267 *the most astonishing*: Moore, *Schrödinger*, 2.

CHAPTER 28. CONFUSION AND THEN UNCERTAINTY

Page

268 *If we are still going to have to*: Moore, *Schrödinger*, 228.

268 *I am convinced that you have made a decisive advance*: Einstein to Schrödinger, 26 April 1926, in Klein, ed., *Letters on Wave Mechanics*, 28.

269 *appears rather mystifying*: Born to Einstein, 15 July 1925, in *Born-Einstein Letters*, 82.

269 *my beloved master*: Schilpp, ed., *Albert Einstein*, 177.

269 *I was from the beginning quite crushed*: Lindley, *Uncertainty*, 90.

269 *he looked like a simple peasant*: Bartel Leendert Van der Waerden, ed., *Sources of Quantum Mechanics* (Amsterdam: North-Holland, 1967), 19.

269 *very quiet and friendly*: Lindley, *Uncertainty*, 90.

270 *right now physics is very confused*: Ibid., 108.

271 *joie de vivre and hope*: Ibid., 116.

271 *based exclusively on relationships*: Ibid., 115.

271 *Heisenberg has*: Einstein to Ehrenfest, 30 September 1920, Folsing, *Albert Einstein*, 566.

271 *the most interesting thing*: Einstein to Besso, 25 December 1925, ibid., 580.

272 *Heisenberg's paper came out*: Mehra, "Satyendra Nath Bose," 141.

272 *more and more I tend*: Einstein to Ehrenfest, 12 February 1926, in Mehra and Rechenberg, *HDQT*, vol. 5, part 2, p. 637.

272 *made much more of an impression*: Interview of Max Born by Archives for the History of Quantum Physics Collection, Niels Bohr Library and Archives, American Institute of Physics, College Park, MD, www.aip.org/history/ohilist /LINK, p. 23. (Cited hereinafter as Born, AHQP interview.)

273 *not such an infernal machine*: Folsing, *Albert Einstein*, 582.

273 *Schrödinger has come out with*: Einstein to Besso, *Einstein Besso Correspondance*, 225.

273 *People were packed*: Mehra and Rechenberg, *HDQT*, vol. 5, part 2, p. 636.

273 *discouraged, if not repelled*: Schrödinger, *Collected Papers on Wave Mechanics*, footnote, p. 46.

273 *what [he] writes about Anschaulichkeit*: Heisenberg to Pauli, 8 June 1926, Moore, *Schrödinger*, 221.

274 *thrown out of the room*: Moore, *Schrödinger*, 222.

274 *we suddenly found him*: Stachel, "Einstein and Bose," 527.

275 *We are all here fascinated*: Einstein to Epstein, 10 June 1926, in Folsing, *Albert Einstein*, 583.

276 *I saw Franck counting particles*: Born, AHQP interview, 25.

276 *I discussed this with him*: Ibid., 26.

277   *Here the whole problem of determinism*: Pais, *Inward Bound*, 257.

277   *the motion of particles follows*: Ibid., 258.

277   *acrimonious debate . . . he believed*: Born, AHQP interview, 25.

277   *I am entirely satisfied*: Born to Einstein, 30 November 1926, Mehra and Rechenberg, *HDQT*, vol. 6, part 1, p. 243. (This letter is not in the published Born-Einstein correspondence.)

277   *Quantum mechanics calls for*: Einstein to Born, 1926, in Folsing, *Albert Einstein*, 585.

278   *isn't that precisely*: Stachel, "Einstein and the Quantum," 388.

278   *possibly I did . . . but it is nonsense. . . . It is the theory*: Ibid.

278   *It must have been one evening*: Stachel, "Einstein and the Quantum," 388.

278   *A good joke* (quoted in footnote): Frank, *Einstein*, 216.

CHAPTER 29: *NICHT DIESE TÖNE*

Page

279   *All the fifty years*: Einstein to Besso, 12 December 1951, in *Einstein Besso Correspondance*, 453.

279   *Here I sit*: Einstein, "Autobiographical Notes," 3.

279   *only a temporary way out*: Ibid., 51.

280   *I know this business*: Pais, *Subtle Is the Lord*, 449.

280   *I am convinced that this theory*: Ibid., 448.

282   *Einstein is therefore clearly involved*: Schilpp, ed., *Albert Einstein*, 174.

283   *the years of searching*: Alice Calaprice, *The Quotable Einstein* (Princeton: Princeton University Press, 1996), 174.

283   *it is my experience*: Einstein, "Autobiographical Notes," 89.

283   *obviously not [a] satisfactory solution*: Mehra and Rechenberg, *HDQT*, vol. 1, part 2, p. 547.

284   *the other day I was*: M. Born to Bohr, 15 January 1925, ibid., 611.

284   *to give our revolutionary efforts*: N. Bohr to R. Fowler, 21 April 1925, in Mehra and Rechenberg, *HDQT*, vol. 1, part 2, p. 613.

284   *fond . . . [which] has a great similarity*: Pais, *Inward Bound*, 259.

285   *Do you really believe*: Pais, *Subtle Is the Lord*, 5.

285   *Physics is an attempt*: Einstein, "Autobiographical Notes," 81.

285   *We have become Antipodean*: Einstein to Born, 7 September 1944, *Born-Einstein Letters*, 146.

285   *nothingness of the hopes*: Einstein, "Autobiographical Notes," 3.

285   *out yonder there was this huge world*: Ibid., 5.

285   *I believe . . . that one of the strongest motives*: Albert Einstein, "Principles of Research," in *Ideas and Opinions*, trans. Sonja Bargmann (New York: Random House, 1954), 225

REFERENCES

EINSTEIN'S WRITINGS AND CORRESPONDENCE

Einstein, Albert. "Autobiographical Notes." In *Albert Einstein: Philosopher-Scientist*, pp. 1–94. Edited by P. A. Schilpp. La Salle: Open Court, 1970.

Einstein, Albert. *The Born-Einstein Letters, 1916–1955: Friendship, Politics and Physics in Uncertain Times*. Translated by Irene Born. New York: MacMillan, 1971.

Einstein, Albert. *The Collected Papers of Albert Einstein*. Translated by Anna Beck and consultation by Don Howard. 12 vols. Princeton: Princeton University Press, 1987–2009. References are to the English translations unless otherwise noted.

Einstein, Albert. *Einstein Besso Correspondance, 1903–1955*. Translated into French by Pierre Speziali. Paris: Hermann, 1972. English translations in the text by the author.

Einstein, Albert. *Ideas and Opinions*. Translated by Sonja Bargmann. New York: Random House, 1954.

BIOGRAPHICAL WORKS ON EINSTEIN

Bernstein, Jeremy. *Albert Einstein*. Edited by Frank Kermode. New York: Penguin Books, 1973.

Calaprice, Alice. *The Quotable Einstein*. Princeton: Princeton University Press, 1996.

D'Amour, Thibault. *Once Upon Einstein*. Translated by Eric Novak. Wellesley: A. K. Peters, 2006.

Dukas, Helen, and Banesh Hoffman, eds. *Albert Einstein: The Human Side*. Princeton: Princeton University Press, 1979.

Folsing, Albrecht. *Albert Einstein: A Biography*. Translated and abridged by Ewald Osers. New York: Penguin Press, 1998.

Frank, Phillip. *Einstein: His Life and Times*. Translated by George Rosen and edited by Schuichi Kusaka. New York: Da Capo Press, 1947.

French, A. P., ed. *Einstein: A Centenary Volume*. Cambridge, MA: Harvard University Press, 1979.

Hentschel, Ann M., and Gerd Grasshoff. *Albert Einstein: Those Happy Bernese Years*. Bern: Staempfli, 2005.

Highfield, Roger, and Paul Carter. *The Private Lives of Albert Einstein*. London: Faber & Faber, 1993.

Hoffmann, Banesh, with the collaboration of Helen Dukas. *Albert Einstein: Creator and Rebel*. New York: Viking Press, 1972.

Isaacson, Walter. *Einstein: His Life and Universe*. New York: Simon & Schuster, 2001.

Levenson, Thomas. *Einstein in Berlin*. New York: Bantam Books, 2003.

Moszkowski, Alexander. *Conversations with Einstein*. Translated by Henry L. Brose. New York: Horizon Press, 1970.

Neffe, Jurgen. *Einstein: A Biography*. Translated by Shelley Frisch. New York: Farrar Strauss Giroux, 2007.

Pais, Abraham. *Einstein Lived Here*. New York: Oxford University Press, 1994.

Pais, Abraham. *Subtle Is the Lord*. Oxford: Oxford University Press, 2005.

Schilpp, P. A., ed., *Albert Einstein: Philosopher-Scientist*. La Salle: Open Court, 1970.

Seelig, Carl. *Albert Einstein: A Documentary Biography*. Translated by Mervyn Savill. London: Staples Press, 1956.

Woolf, Harry, ed. *Some Strangeness in Proportion: Einstein Centennial*. Reading: Addison-Wesley, 1980.

EINSTEIN AND QUANTUM THEORY

Bolles, Edmund Blair. *Einstein Defiant: Genius versus Genius in the Quantum Revolution*. Washington, DC: John Henry Press, 2005.

Klein, Martin J. "Einstein and Wave-Particle Duality." *The Natural Philosopher*, vol. 3, 1964, pp. 1–49.

Stachel, John. "Einstein and the Quantum" and "Bose and Einstein." In *Einstein from B to Z*, vol. 9, pp. 367–444. Edited by Don Howard. Boston: Birkhauser, 2002.

QUANTUM THEORY AND QUANTUM MECHANICS

Haar, D. Ter. *The Old Quantum Theory*. Oxford: Pergamon Press, 1967.

Hermann, Armin. *The Genesis of Quantum Theory (1899–1913)*. Cambridge, MA: MIT Press, 1971.

Kuhn, Thomas S. *Black-Body Theory and the Quantum Discontinuity, 1894–1912*. Chicago: University of Chicago Press, 1978.

Lindley, David. *Uncertainty: Einstein, Bohr, and the Struggle for the Soul of Science*. New York: Doubleday, 2007.

Mehra, Jagdish, and Helmut Rechenberg. *The Historical Development of Quantum Theory*, vols. 1–5. New York: Springer-Verlag, 1982–1987.

Pais, Abraham. *Inward Bound: Of Matter and Forces in the Physical World*. New York: Clarendon Press, 1986.

Van der Waerden, Bartel Leendert, ed. *Sources of Quantum Mechanics*. Amsterdam: North-Holland, 1967.

Wheaton, B. R. *The Tiger and the Shark: Empirical Roots of Wave-Particle Dualism*. Cambridge: Cambridge University Press, 1983.

BIOGRAPHICAL MATERIAL ON OTHER SCIENTISTS

Abragam, A. "Louis De Broglie." *Biographical Memoirs of Fellows of the Royal Society*, vol. 34, 1988, pp. 22–41.

AHQP Interviews of Louis De Broglie, by T. S. Kuhn, A. George, and T. Kahan, on January 7 and 14, 1963. Archives for the History of Quantum Physics Collection, Niels Bohr Library and Archives, American Institute of Physics, College Park, MD, www.aip.org/history/ohilist/LINK.

AHQP Interview of Max Born, by T. S. Kuhn and F. Hund on October 17, 1962. Archives for the History of Quantum Physics Collection, Niels Bohr Library and Archives, American Institute of Physics, College Park, MD, www.aip.org/history/ohilist/LINK.

Barkan, Diana Kormos. *Walther Nernst and the Transition to Modern Physical Science*. Cambridge: Cambridge University Press, 1999.

Barut, Asim O., Alwyn van der Merwe, and Jean-Pierre Vigier, eds. *Quantum Space and Time—the Quest Continues: Studies and Essays in Honour of Louis De Broglie, Paul Dirac and Eugene Wigner*. Cambridge: Cambridge University Press, 1984.

Blanpied, W. "Satyendranath Bose: Co-founder of Quantum Statistics." *American Journal of Physics*, September 1972, pp. 1212–1220.

Coffey, Patrick. *Cathedrals of Science*. Oxford: Oxford University Press, 2008.

Crawford, Elisabeth. "Arrhenius, the Atomic Hypothesis, and the 1908 Nobel Prizes in Physics and Chemistry." *Isis*, vol. 75, 1984, pp. 503–22.

Cropper, William. *Great Physicists: The Life and Times of Leading Physicists from Galileo to Hawking*. Oxford: Oxford University Press, 2004.

Crowther, James Gerald. *Scientific Types*. New York: Dufour, 1970.

Duck, Ian, and E.C.G. Sudarshan, eds. *100 Years of Planck's Quantum*. Singapore: World Scientific Publishing, 2000.

Heilbron, J. L. *Dilemmas of an Upright Man: Max Planck as Spokesman for German Science*. Berkeley: University of California Press, 1986.

Heisenberg, Werner. *Encounters with Einstein: And Other Essays on People, Places, and Particles*. Princeton: Princeton University Press, 1989.

Heisenberg, Werner. *Physics and Philosophy: The Revolution in Modern Science*. World Perspectives, vol. 19. Edited by Ruth Nanda Anshen. New York: Harper & Brothers, 1958.

Klein, Martin J., ed. *Letters on Wave Mechanics*. New York: Philosophical Library, 1967.

Klein, Martin J. *Paul Ehrenfest: The Making of a Theoretical Physicist*, vol. 1. Amsterdam: North-Holland, 1970.

Lorentz, H. A. *Impressions of His Life and Work*. Edited by G. L. de Haas-Lorentz. Amsterdam: North-Holland , 1957.

Kragh, Helge S., *Dirac: A Scientific Biography*. Cambridge: Cambridge University Press, 1990.

Marage, Pierre, and Grégoire Wallenborn, eds. *The Solvay Councils and the Birth of Modern Physics*. Science Networks, vol. 22. Basel: Birkauser Verlag, 1999.

Maxwell, J. C. *The Scientific Papers of James Clerk Maxwell*, vol. 2. Edited by W. D. Niven. Dover, NY: Dover Publications, 1965.

Maxwell, James Clerk. "Molecules." *Nature*, September 1873, pp. 437–441, Victorian Web, http://www.victorianweb.org/science/maxwell/molecules.html, accessed July 20, 2008.

Mehra, Jagdish. "Satyendra Nath Bose." *Biographical Memoirs of Fellows of the Royal Society*, vol. 21, 1975, pp. 117–154.

Mendelssohn, K. *The World of Walther Nernst: The Rise and Fall of German Science, 1864–1941*. Pittsburgh: University of Pittsburgh Press, 1973.

Moore, Walter. *Schrödinger: Life and Thought*. Cambridge: University of Cambridge Press, 1989.

Nagel, Bengt. "The Discussion Concerning the Nobel Prize for Max Planck." In *Science, Technology and Society in the Time of Alfred Nobel: Nobel Symposium 52*. Edited by C. G. Bernhard, E. Crawford, and P. Sorbom. New York: Pergamon Press, 1982.

Pais, Abraham. *Subtle Is the Lord*. Oxford: Oxford University Press, 2005.

Planck, Max. *Scientific Biography and Other Papers, 1949*. Translated by Frank Gaynor. New York: Philosophical Library, 1949.

Schrödinger, Erwin. *What Is Life? The Physical Aspect of the Living Cell with Mind and Matter & Autobiographical Sketches*. Cambridge: Cambridge University Press, 1958.

Scott, William T. *Erwin Schrödinger: An Introduction to His Writings*. Amherst: University of Massachusetts Press, 1967.

Stachel, John. "Einstein and Bose." In *Einstein from B to Z*, vol. 9, pp. 519–538. Edited by John Stachel and Don Howard. Boston: Birkhauser, 2002.

Strutt, Robert John. *The Life of Lord Rayleigh*. Madison: University of Wisconsin Press, 1968.

Tolstoy, Ivan. *James Clerk Maxwell*. Edinburgh: Canongate, 1981.

Wali, K. "The Man behind Bose Statistics." *Physics Today*, October 2006, p. 46.

ORIGINAL SCIENTIFIC RESEARCH ARTICLES (CHRONOLOGICAL)

1. Max Planck, "On an Improvement of Wien's Equation for the Spectrum," *Proceedings of the German Physical Society*, vol. 2, p. 202 (1900); reprinted in translation in Haar, *The Old Quantum Theory*, 79–81.

2. Max Planck, "On the Theory of the Energy Distribution Law of the Normal Spectrum," *Proceedings of the German Physical Society*, vol. 2, p. 237 (1900); reprinted in translation in Haar, *The Old Quantum Theory*, 82–90.

3. Lord Rayleigh, "Remarks upon the Law of Complete Radiation," *Philosophical Magazine*, vol. 49, pp. 539–40 (1900); reprinted in *Scientific Papers by Lord Rayleigh*, vol. 6, doc. 260, pp. 483–485, Dover, New York (1964).

4. Lord Rayleigh, "The Law of Partition of Kinetic Energy," *Philosophical Magazine*, vol. 49, pp. 98–118 (1900); reprinted in *Scientific Papers by Lord Rayleigh*, vol. 6, doc. 253, pp. 433–451, Dover, New York (1964).

5. Albert Einstein, "On the General Molecular Theory of Heat," *Annalen der Physik*, vol. 14, pp. 354–362 (1904); reprinted in *CPAE*, vol. 2, doc. 5, pp. 68–77.

6. Albert Einstein, "On a Heuristic Point of View concerning the Production and Transformation of Light," *Annalen der Physik*, vol. 17, pp. 132–148 (1905); reprinted in *CPAE*, vol. 2, doc. 14, pp. 86–103.

7. Albert Einstein, "On the Electrodynamics of Moving Bodies," *Annalen der Physik*, vol. 17, pp. 891–921 (1905); reprinted in *CPAE*, vol. 2, doc. 23, pp. 140–171.

8. Albert Einstein, "On the Theory of Light Production and Light Absorption," *Annalen der Physik*, vol. 20, p. 199 (1906); reprinted in *CPAE*, vol. 2, doc. 34, pp. 192–199.

9. Albert Einstein, "Planck's Theory of Radiation and the Theory of Specific Heat," *Annalen der Physik*, vol. 22, pp. 180–190 (1907); reprinted in *CPAE*, vol. 2, doc. 38, pp. 214–224.

10. Albert Einstein, "On the Present Status of the Radiation Problem," *Physikalische Zeitschrift*, vol. 10, pp. 185–193 (1909); reprinted in *CPAE*, vol. 2, doc. 56, pp. 357–375.

11. Albert Einstein, "On the Development of Our Views concerning the Nature and Constitution of Radiation," *Physikalische Zeitschrift*, vol. 10, pp. 817–826 (1909), presented at the 81st Meeting of the German Scientists and Physicians, Salzburg, September 21, 1909; reprinted in *CPAE*, vol. 2, doc. 60, pp. 379–394.

12. "Discussion Following the Lecture: On the Development of Our Views concerning the Nature and Constitution of Radiation," *Physikalische Zeitschrift*, vol. 10, pp. 825–826 (1909), presented at the 81st Meeting of the German Scientists and Physicians, September 21, 1909; reprinted in *CPAE*, vol. 2, doc. 61, pp. 395–398.

13. Albert Einstein, "On the Present State of the Problem of Specific Heats," *Proceedings of the Solvay Conference*, October 30–November 3, 1911; reprinted in *CPAE*, vol. 2, doc. 26, pp. 419–420.

14. Niels Bohr, "On the Constitution of Atoms and Molecules," *Philosophical Magazine*, vol. 26, p. 1 (1913); reprinted in *The Old Quantum Theory*, by D. Ter Haar, pp. 132–159.

15. Albert Einstein, "Emission and Absorption of Radiation in Quantum Theory," *Proceedings of the German Physical Society*, vol. 18, pp. 318–323 (1916); reprinted in *CPAE*, vol. 6, doc. 34, pp. 212–216.

16. Albert Einstein, "On the Quantum Theory of Radiation," *Physikalische Gesellschaft Zurich, Mitteilungen*, vol. 18 (1916); reprinted in *CPAE*, vol. 6, doc. 38, pp. 220–233.

17. Albert Einstein, "On the Quantum Theorem of Sommerfeld and Epstein," *Proceedings of the German Physical Society*, vol. 19 (1917); reprinted in *CPAE*, vol. 6, doc. 45, pp. 434–443.

18. S. N. Bose, "Planck's Law and the Light Quantum Hypothesis," *Zeitschrift für Physik*, vol. 26, p. 178 (1924); reprinted in O. Theimer and B. Ram, "The Beginning of Quantum Statistics," *American. Journal of Physics*, vol. 44, pp. 1056–1057 (1976).

19. S. N. Bose, "Thermal Equilibrium in the Radiation Field in the Presence of Matter," *Zeitschrift für Physik*, vol. 27, p. 384 (1924); reprinted in O. Theimer and

B. Ram, "Bose's Second Paper: A Conflict with Einstein," *American Journal of Physics*, vol. 45, pp. 242–246 (1976).

20. Albert Einstein, "Quantum Theory of the Monatomic Ideal Gas," *Proceedings of the Prussian Academy of Sciences*, vol. 22, p. 261 (1924); reprinted in translation in I. Duck and E.C.G. Sudarshan, eds., *Pauli and the Spin-Statistics Theorem*, World Scientific, Singapore (1997), 82–87.

21. Albert Einstein, "Quantum Theory of the Monatomic Ideal Gas, Part Two," *Proceedings of the Prussian Academy of Sciences*, vol. 1, p. 3 (1925); reprinted in translation in I. Duck and E.C.G. Sudarshan, eds., *Pauli and the Spin-Statistics Theorem*, World Scientific, Singapore (1997), 89–99.

22. Albert Einstein, "On the Quantum Theory of the Ideal Gas," *Proceedings of the Prussian Academy of Sciences*, vol. 3, p. 18 (1925); reprinted in translation in I. Duck and E.C.G. Sudarshan, eds., *Pauli and the Spin-Statistics Theorem*, World Scientific, Singapore (1997), 100–107.

23. Louis de Broglie, "Black Radiation and Light Quanta," *Journal de Physique et le Radium*, vol. 3, p. 422 (1922); reprinted in *Selected Papers on Wave Mechanics by Louis de Broglie and Leon Brillouin*, vols. 1–7, Blackie and Sons, London (1928).

24. Louis de Broglie, "A Tentative Theory of Light Quanta," excerpt from *Philosophical Magazine*, vol. 47, p. 446 (1924); reprinted in I. Duck and E.C.G. Sudarshan, eds., *100 Years of Planck's Quanta*, chapter 4, World Scientific, Singapore (2000), 128–141.

25. Louis de Broglie, "Studies on the Theory of Quanta," PhD thesis, originally published in *Annales de Physique*, vol. 3, p. 22 (1925).

26. Erwin Schrödinger collected nine of his seminal papers on the wave equation into a volume titled *Abhandlungen der Wellenmechanik* (Treatise on Wave Mechanics), which was originally published in 1927. These papers are available in English translation in E. Schrödinger, *Collected Papers on Wave Mechanics*, Chelsea Publishing, New York (1978). The nine papers are titled "Quantisation as a Problem of Proper Values, Parts I, II, III, IV," "The Continuous Transition from Micro- to Macro-Mechanics," "On the Relation between the Quantum Mechanics of Heisenberg, Born and Jordan, and That of Schrödinger," "The Compton Effect," "The Energy-Momentum Theorem for Material Waves," and "The Exchange of Energy According to Wave Mechanics." Note that the term "proper value" was the chosen translation for the German term *Eigenvalue*, which has become standard mathematical terminology in English as well.

INDEX

Note: page numbers in italics refer to images.